Democracy Disconnected

Why is dissatisfaction with local democracy endemic, despite the spread of new participatory institutions? This book argues that a key reason is the limited power of elected local officials, especially to produce the City. City Hall lacks control over key aspects of city decision-making, especially under conditions of economic globalisation and rapid urbanisation in the urban South.

Demonstrated through case studies of daily politics in Hout Bay, *Democracy Disconnected* shows how Cape Town residents engage local rule. In the absence of democratic control, urban rule in the Global South becomes a complex and contingent framework of multiple and multilevel forms of urban governance (FUG) that involve City Hall, but are not directed by it. Bureaucratic governance coexists alongside market, developmental and informal forms of governance. This disconnect of democracy from urban governance segregates people spatially, socially, but also politically. Thus, while the residents of Hout Bay may live next to each other, they do not live with each other.

This book will be a valuable resource for students on programmes such as urban studies, political science, sociology, development studies, and political geography.

Fiona Anciano is a Senior Lecturer in the Department of Political Studies at the University of the Western Cape, South Africa. Her research areas include civil society, democracy, and urban politics.

Laurence Piper is Professor of Political Studies at the University of the Western Cape, South Africa and visiting Professor at the University West, Sweden. His research area is urban politics in the Global South.

Routledge Studies in Urbanism and the City

This series offers a forum for original and innovative research that engages with key debates and concepts in the field. Titles within the series range from empirical investigations to theoretical engagements, offering international perspectives and multidisciplinary dialogues across the social sciences and humanities, from urban studies, planning, geography, geohumanities, sociology, politics, the arts, cultural studies, philosophy, and literature.

Rebel Streets and the Informal Economy
Street Trade and the Law
Edited by Alison Brown

Mega-events and Urban Image Construction
Beijing and Rio de Janeiro
Anne-Marie Broudehoux

Urban Geopolitics
Rethinking Planning in Contested Cities
Edited by Jonathan Rokem and Camillo Boano

Contested Markets, Contested Cities
Gentrification and Urban Justice in Retail Spaces
Edited by Sara González

The City as a Global Political Actor
Edited by Stijn Oosterlynck, Luce Beeckmans, David Bassens, Ben Derudder, Barbara Segaert and Luc Braeckmans

Democracy Disconnected
Participation and Governance in a City of the South
Fiona Anciano and Laurence Piper

For more information about this series, please visit: www.routledge.com/series/RSUC

Democracy Disconnected

Participation and Governance in a City of the South

Fiona Anciano and Laurence Piper

Routledge
Taylor & Francis Group

LONDON AND NEW YORK

First published 2019 by Routledge

2 Park Square, Milton Park, Abingdon, Oxfordshire OX14 4RN

52 Vanderbilt Avenue, New York, NY 10017

Routledge is an imprint of the Taylor & Francis Group, an informa business

First issued in paperback 2019

British Library Cataloguing-in-Publication Data
A catalogue record for this book is available from the British Library

Library of Congress Cataloging-in-Publication Data
A catalog record has been requested for this book

ISBN: 978-1-138-54105-4(hbk)
ISBN: 978-0-367-28085-7(pbk)

Typeset in Times New Roman
by Out of House Publishing

Contents

Illustrations

Figures

Maps

Tables

Acknowledgements

We live in a country steeped in a history of inequality and violence, and yet it is populated with resilient citizens who are determined to build a better future. It is these everyday South Africans who inspired us to understand how they navigate their life in the Global South, and to write this book.

This book was a truly collaborative effort. We are following the UK convention of alphabetic order of equal co-authors as the book was published in the UK. Additionally, many people and organisations over the years have helped us along the way. We would like to thank David Southwood for his excellent photographs, Orli Setton for the expert help with images, and Sherran Clarence for her careful proofreading. Intellectually, we owe a debt to Lawrence Hamilton, Wayne Hugo and Federico Rossi who gave us comments on the drafts of our work, and especially to Andries du Toit for his close and critical reading.

We would not have got started without Joanna Wheeler, Rory Liedeman, Anthony Muteti, and Thuli Ntshingila, and we would not have finished without the work of Roslyn Bristow, Conrad Meyer, and Zikhona Sikota. A special thanks to PASSOP for introducing us to Hout Bay, the Sustainable Livelihoods Foundation for their great survey work in Imizamo Yethu and the research on taxis. Thanks also to Actuality for their collaboration on the water and waste chapter and the leader of Thrive for giving of her time so generously.

This project was only possible because of the generosity of spirit of the people of Hout Bay, literally hundreds of whom assisted us directly and indirectly over the years. Andrew, Angelo, Ashley, Bernard, Brian, Bronwyn, Claire, Clifford, Cormac, Gerda, Greg, Jan, Johan, JJ, Khaya, Kiara, Kenny, Lelethu, Liz, Mark, the Pastor, Samkelo, Shannon and Vincent deserve a special mention. We would also like to thank the officials and politicians in the City of Cape Town and the Western Cape Provincial government who spent many hours answering our questions.

There are several research organisations that have, over the years, supported the work that contributed to this project. These include the ESRC project on Agency and Governance in Contexts of Civil Conflict (RES-167-25-0481), the Participedia project out of the University of British Columbia, but most of all the National Research Foundation (NRF) of South Africa, who provided a grant for a project on Understanding Urban Violence (no: 87824) and a Thuthuka Rating Track grant (no: 106923).

We are privileged to work with a wonderful team in the Department of Political Studies at the University of the Western Cape. Thank you for being true colleagues. Thanks also to our friendly and talented colleagues at University West Sweden, for hosting Laurence during the final months of writing the book. A special thanks to Professor Per Assmo for making it all possible.

Most of all, thank you to our supportive families; to Verna and Pete who instilled the belief one can achieve whatever they set their mind to, and to Damien, Ava, and Ella for supporting, with love, these pursuits. Lastly, a special thanks to Ethan, Samuel, and Sherran for coming halfway around the world in support of this book.

Fiona Anciano, Cape Town, South Africa
Laurence Piper, Trollhättan, Sweden

Abbreviations

ACDP	African Christian Democratic Party
ANC	African National Congress
BRT	Bus Rapid Transport System
CATA	Cape Amalgamated Taxi Association
CBD	Central Business District
CCP	Community Crime Prevention
CCT	City of Cape Town
CTZS	Cape Town Zoning Scheme
CUTA	Cape United Taxi Association
DA	Democratic Alliance
DAFF	Department of Agriculture, Fisheries and Forestry
DAG	Development Action Group
DFA	Development Facilitation Act of 1995
EFF	Economic Freedom Fighters
EPWP	Extended Public Works Programme
FF+	Freedom Front Plus
FUG	Framework of Urban Governance
HBCA	Hout Bay Civic Association
HBP	Hout Bay Partnership
HBRCF	Hout Bay Rivers Catchment Forum
HBRRA	Hout Bay Residents and Ratepayers Association
HiDA	Hangberg in situ Development Association
IDP	Integrated Development Plan
NGO	Non-Governmental Organisation
NNP	New National Party
PMF	Peace and Mediation Forum
PPP	Public-Private Partnership
PR	Proportional Representation
PRA	Participatory Rural Appraisal
SANCO	South African National Civics Organisation
SANParks	South African National Parks
SAPS	South African Police Services

SDF	Spatial Development Framework
SLF	Sustainable Livelihoods Foundation
SSF Policy	Small-scale Fishing Policy
SPLUMA	Spatial Planning and Land Use Management Act of 2013
UN	United Nations
UISP	Upgrading of Informal Settlements Programme

Introduction

The paradox: more participation but an enduring democratic deficit

Liberal democracy is in trouble. Authoritarian regimes from China to Turkey are gaining influence on the world stage, right-wing nationalism is on the rise globally, and developing countries have appeared to stall en route to full democracy. Perhaps even more parlous for liberal democracy than the rising power of its opponents, is the declining satisfaction of its supporters. Pippa Norris (2011) argues that, while demand for democracy is at an all-time high, satisfaction with the supply of democracy by government is dropping. This difference between citizens' 'demand' for democratic institutions, values, and principles and the 'supply' from government is termed a 'democratic deficit', and is clearly manifested in recent waves of protest and populist politics across all continents.

An important response to the liberal democracy deficit has been the rise of new forms of participatory democracy (Gaventa 2006; Wampler 2012). This idea echoes Northern debates on deliberative democracy (Cohen 1997), but is mostly a Southern invention from radical political parties and social movements, largely urban-based, and often driven from the peripheries of the city. Hence, the origins of the archetypal participatory democratic institution, participatory budgeting, lie in the rise of the *Partido dos Trabalhadores* (Worker's Party) in Porto Alegre, Brazil. This city is also home to the World Social Forum and the attendant demand of the 'right to the city' championed by social movements. Harvey (2012: xii) argues that central to the idea of the 'right to the city' is a desire by peripheral residents to claim control over daily urban life. Related to this are notions of active citizenship and, especially in Brazil, the 'insurgent citizenship' developed in the *favelas* (slums) that strive to redefine governance in more inclusive and egalitarian directions (Holston 2008).

These ideas of participatory democracy, the right to the city and insurgent citizenship resonate across the Global South and beyond, and are evident in a range of institutional innovations and democratic experimentation. While some of this relates to rural development work, such as Robert Chambers' (1992) Participatory Rural Appraisal (PRA), most is focused on the urban. In addition to tracking rapid urbanisation across the Global South, the rise of participatory democracy and insurgent citizenship also parallel the growing

Figure I.1 Public participation in Hangberg, Hout Bay
(Photo by Macherez Yann)

inequality of urban life, in particular the contrast between the burgeoning slums of the poor and the gated communities of the wealthy (Davis 2006). Indeed insurgent citizenship and family conceptions like 'spectacular politics' (Robins 2014), 'occupancy urbanisms' (Benjamin 2008) or 'subaltern urbanisms' (Roy 2011) are associated with the peripheral, informal and 'grey' spaces (Yiftachel 2009) of the city.

Yet a paradox endures. The rise of participatory democracy, social movement politics, and insurgent citizenship has not reduced the levels of democratic deficit that Norris (2011) has tracked for over a decade. Indeed, according to her, levels of democratic satisfaction remain largely unchanged in states like Brazil where participatory democracy is relatively advanced. While dissatisfaction is often linked to the 2008 global economic crisis, research confirms that it directly includes frustration with democracy itself. As Ortiz, Burke, Berrada, & Cortès (2013: 6) point out:

> the most sobering finding ... is the overwhelming demand ... not for economic justice per se, but for what prevents economic issues from being addressed: a lack of 'real democracy', which is a result of people's growing awareness that policy-making has not prioritized them – even when it has claimed to – and frustration with politics as usual and a lack of trust in the existing political actors, left and right.

At the same time, though, advocates of participatory democracy have pronounced on its growing success (The Hunger Project & UN Democracy Fund 2014). There is over 10 years of research confirming the positive real-world impact of participatory democracy including participatory budgeting (Baiocchi, Heller, & Silva 2011; Gonçalves 2013), health councils (Coelho 2006), and even participatory impacts on national policy (Pogrebinschi & Samuels 2014). This is further to the extensive and often profound impact of direct resident participation in development projects on citizenship, participatory practices, state accountability, and social cohesion (Gaventa & Barrett 2010).

What are we to make of this paradox? How can participatory democracy be working and yet satisfaction with national democracy remain low? Granted, cities are not states, and Norris relies on World Survey data that specifically asks respondents for their views of democracy at a country level. Are we then comparing apples and oranges? Perhaps participatory democracy is enhancing popular perceptions of local democracy while not influencing national perceptions? While there is some evidence to support this claim (Johnson 2015), we cannot explain away this paradox exclusively in terms of the distinction between national and local democracy. The main reason for this, as Pierre (1995: 38) points out, is that local government has a dual role, on the one hand 'to act as the vehicle of local democracy, providing services responsive to local needs and conditions', and on the other hand to 'constitute the local branch of the nation-state's administrative apparatus, executing state policies in key-policy areas'.

Further, it is a global trend that citizens participate less in elections for local government than for national, and see local government as a junior partner in a larger system of government (Morlan 1984; Frandsen 2002; Hajnal & Lewis 2003). Given this integration of local government into the national state in practice and in perception, it makes sense to assume that, when asked about the health of a country's democracy, as in the World Values Survey that Norris draws upon, respondents will indicate a general impression of the system of government as a whole, including local government. Thus, we can treat satisfaction with democracy as a general assessment of a complex system of democratic governance from the national to the local level.

If assessments of local democracy will always figure in satisfaction with democracy in general, can we conclude then, that participatory democracy is not making a significant difference to country-level assessments of the supply of democracy? The main argument here is that the new institutions of participatory democracy are simply not widespread enough to impact on popular perceptions of democratic governance. In this regard, Sónia Gonçalves (2013: 96) estimated that in 2004 '30 percent of the Brazilian population lived in municipalities that used participatory budgeting as a means of deciding the allocation of local resources'. Further, while it is found across the Global South, participatory democracy exists more as the exception than the rule.

Often associated with development projects, or limited to sections of some cities, participatory democracy is simply not the norm.

There is clearly something to this argument. Despite this, we still hold that there is more to our paradox than limited implementation of participatory democracy, for, even in those countries and contexts where it is well implemented, satisfaction with democracy remains low (Norris 2011). We argue that participatory democracy can, and is, making a positive difference to local democracy across the Global South, but also that, for most citizens, the promised changes in governance are not being fulfilled. Further, we suggest an important reason for this is that local democracy is disconnected from, or only partially connected to, many of the important forms of governance that shape urban rule. Simply put, because key forms of governance over the city are beyond the control of elected representatives, democratic institutions have little means of keeping those who govern accountable.

There are further challenges for democratic rule associated with the inability of democratic institutions to hold certain forms of governance accountable. Central here are the undemocratic logics of certain forms of governance, and the fact that multiple and contending forms of governance coexist in spatially contingent ways that might segregate and disorder urban systems of rule. This we term a topological (Collier 2009) framework of multiple and contending forms of urban governance (FUG). From the perspective of urban residents, especially the urban poor, this creates a frustrating situation where growing democratic opportunities offer limited to no influence over the multiple forms of governance and diverse sets of authorities who decide how they must live. The consequence is that, from a governance point of view, democratic citizenship exists episodically around elections and participation in public forums, but in daily existence, most of the time, residents are treated like consumers of services, clients of the state, or are marginalised from the formal system altogether. Spatially, too, different suburbs are governed in different ways with the wealthy relying on market means, while the poor wait for developmental projects or meet their livelihood and shelter needs themselves.

The mainstream model of urban democratic rule

In making this case we begin with the paradox of more participation but less satisfaction with democracy. To refine this problem into a set of research questions, we first outline the mainstream assumptions of urban democratic rule before identifying the key challenges confronting this model, drawn from current literature on the urban South.

In general terms, liberal democratic rule is imagined as a virtuous feedback circle between politics, democracy, and government (see Figure I.2). This model is an ideal-type from which various instances will diverge to some extent. However, it is useful nevertheless as it is developed from existing forms in the North spread around the world by colonialism, and the later political ideas exported globally after the end of the Cold War under *pax Americana*.

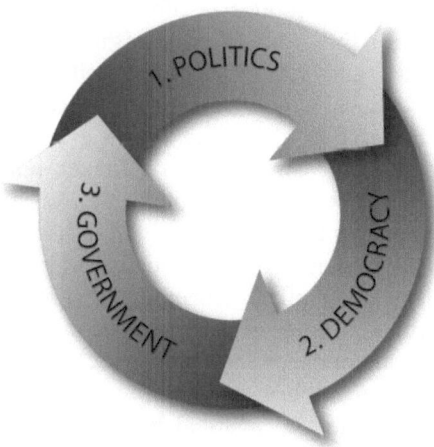

Figure I.2 Model of democratic rule

Consequently, it is reasonable to expect a strong resonance between the model and the forms of city rule in most democracies around the world.

The first of the three components of this model, politics, is conceived of in terms of individuals or groups, equal in principle and in law, forming critical views on issues that they identify as important for rulers to address, and organising and mobilising around these (Dahl 1989). A core assumption of liberal politics is that different people will hold contending views to those of rulers, or to each other, and should be able to express these in public. For liberals this process of politicisation requires the guarantee of certain individual rights to speech as well as free press, academic freedom, and freedom of religion. Also important are rights of association and movement, and the mobilisation of groups to contest office or influence policy is further protected by rights to assembly and peaceful protest (Mill 1859).

Key to linking political issues and groups to government are political parties that help define collective interests, educate citizens, shape group identities, recruit future leaders, develop policies, organise the contestation of power, and help legitimate the political system (Dalton & Wattenberg 2000). Also important are civil society organisations for, although they usually do not contest office, they assist in organising citizens, developing and lobbying for policy positions, recruiting and training leaders, and building democratic practices among citizens (Almond & Verba 1989; Putnam, Leonardi, & Nanetti 1994). In short, liberal politics is based on a set of rules that are intended to allow any individual or group to politicise an issue, and organise and mobilise around it.

The second component, democracy, is understood primarily in terms of institutions through which individuals and groups can contest the offices of rule, or influence the decisions of rulers, in ways that uphold the values of equality, freedom and fraternity. In terms of contesting office, not only is any citizen entitled to run for office, but also the choice of official is made through free, fair, and frequent elections. As a result, electoral outcomes reflect the real choice of the majority of voters (Dahl 1973, 1989). As already noted, political parties are usually central to contests for office, and almost all rulers are leaders of political parties. In a democracy, any group may form a political party and membership is open to all citizens. Critical to all of this is a free, diverse, and independent press that offers multiple means through which to report, examine, and comment on the decisions of rulers (Dahl 1973, 1989).

Furthermore, in addition to contesting office, citizens and groups may seek to influence office-bearers through a range of formal mechanisms from legally required consultations on draft budgets, laws, and policies, through representative forums such as citizen assemblies, to communication one-on-one with office-bearers through, for example, personal lobbying. In many parts of the Global South, new institutions of participatory democracy have been introduced to formalise citizen influence over decision-making by rulers between elections, such as participatory budgeting or requirements for consultation on development projects (Cornwall & Coelho 2007).

Lastly, in the mainstream model, the making and implementing of decisions is enacted by government, led by elected officials 'in that they are not subject to the tutelary control of military or clerical leaders' (Levitsky & Way 2002: 53). Thus, a key feature of democratic rule is that elected leaders do actually make the important decisions. These decisions come in three common forms. First, rulers must deal with events that confront the state, for example whether to bid for the Olympics or to retaliate to a military incursion. Second, they must develop policies or plans of action for the economy, education, health, security, and so on. Third, they must introduce new laws to enable these decisions and policies, subject to some form of judicial or constitutional oversight (Dahl 1989).

While the capacity of representatives to govern is an essential element of the model of contemporary liberal democratic rule, it is also deliberately limited. This limitation exists both in terms of the framework of law, especially the constitution, and in the separation of powers within the state to prevent the abuse of power by rulers (Hamilton, James, & Jay 1961). At the heart of the mainstream model then is a balance between constituting a government that reflects the choice of the majority of citizens, and empowering it to act, but not in ways that threaten individual and group rights that make democracy possible. Indeed, should government make illegal choices then it can be challenged through the law, and should it make unpopular choices, it can be changed at the next election. This self-correcting capacity aims to make liberal democratic rule a virtuous circle of politics, democracy, and government.

In respect of *local* democratic rule, the same model of democracy applies but with an important qualification. This is that local government is never self-governing but always part of a larger state system. This opens the possibility that City Hall may take positions on issues that conflict with the positions of the national government. Depending on the design of the system, one level (usually national) will prevail over the other. The existence of conflict between national and local is thus not evidence of a lack of democracy in and of itself. It is possible for both levels of state to represent the views of the majority of citizens in both of their domains authentically, and for their divergent groups of citizens to hold different views. Hence, the mere fact that one level of the state can impose its will on another is not intrinsically undemocratic.

The realities confronting urban democratic rule in the Global South

Having outlined the mainstream model of urban democratic rule, we identify some key challenges in the Global South that confront the idea of a virtuous circle between politics, democracy, and government. This is important to do because it alerts us to possible reasons why local democracy is disappointing residents of cities of the South. Before doing this it is useful to sketch some general features of the urban South that mark it as distinct from the North.

Perhaps the most widely observed point about the urban South since the end of the Cold War is the fact of rapid urbanisation. From 1950 to 2015, the urban population of the world increased 500 per cent from 746 million to 4.9 billion, and at some point in 2007, the world's population become more urban than rural for the first time in human history. By 2014, 54 per cent of the globe was urban, and this number is expected to grow to 66 per cent by 2050. Critically, most of this growth has been in the urban South, and especially the smaller cities of less than 500,000 people. By 2050, 90 per cent of the world's urban population will live in Africa and Asia (UN 2014: 1).

A key challenge associated with rapid urbanisation in the Global South is that migration has outstripped the growth in urban jobs, with a few exceptions, such as China. According to Davis (2006: 15), this is in part due to global neo-liberal deregulation of the agricultural sector that has created 'surplus' rural labour who now move to the cities for new opportunities. Consequently, urbanisation has brought with it the growth of slums, although as UN Habitat (2016) points out, slums were down from 39 per cent of the urban South in 2000 to 30 per cent in 2016. Alongside informal settlement in slums come informal ways of securing an income, through street trade, retailing food or alcohol, manufacturing components at the start of a commodity value chain, providing services such as hair care, sex work, and so on. Much of this is without the required licences, outside of designated trading zones, and not to industry standards (Ledeneva 2018). In addition, many migrants cross national boundaries to seek economic opportunities, and some are asylum seekers from political or social persecution.

In addition to reshaping human settlement and livelihoods, urbanisation in the Global South profoundly affects the environment too. This is because those who live in urban areas have very different consumption patterns than rural residents (Parikh et al. 1991). The result is fresh water scarcity, deforestation, air and water pollution, much of which results in climate change. Indeed, meeting the challenge of climate change globally in many ways requires addressing new forms of urban rule (Hughes 2017). Lastly, the rapid pace of social change brought about by urbanisation in the Global South has not been met with an equally rapid and appropriate policy response. As Watson (2009: 2259) points out, much planning theory is based on Northern assumptions that do not accommodate the reality of 'the problems of poverty, inequality, informality, rapid urbanisation and spatial fragmentation particularly (but not only) in cities of the Global South'. Key from a policy perspective, she concludes, is recognition of 'increasingly marginalised urban populations surviving largely under conditions of informality'.

As we shall argue in what follows, the failure to include the (often migrant) poor in formal systems of urban rule reflects the lack of productive power held by City Hall, indeed any formal or informal authority to create the urban South. This quest for productive power, we suggest, is the key challenge that will confront rule in the urban South into the future. Against this background of the distinctive but general features of life in the urban South, we now focus on what we take to be key empirical realities that pose challenges for the virtuous circle of liberal democratic rule outlined above.

Politics and the exclusion of the urban poor

In terms of politics, the key challenge to the mainstream model of democratic rule in the urban South is one of exclusion from democratic citizenship. This means that, either formally or informally, residents of the urban South are overtly excluded from, or find it difficult to participate in, activities to politicise an issue in the public domain, and organise or mobilise around it. Three kinds of exclusion from democratic citizenship are commonly observed: the exclusion of the foreigner, both legally and in terms of belonging; the exclusion of women and minority groups who enjoy legal standing but lower social status; and the exclusion of the urban poor who may belong to the city but cannot live fully by its rules. Of these, the latter is by far the most commonly observed, and it is widely remarked upon how the politics of the urban poor is exceptional to the model of liberal democratic rule.

To begin with a more positive account of the exclusion of the urban poor, we start with Holston's (2008) account of insurgent citizenship in Brazil. Reflecting on shifts in politics over a generation, following the advent of democracy in 1988, Holston identifies the emergence of an insurgent citizenship from the peripheries of the city that, in the name of democratic equality, challenges the traditional exclusion of the poor from political life. This he frames as a tactically disruptive but ideologically positive politics originating among residents of *favelas* (slums), who literally make their own life in the city by building their own homes and making new livelihoods, and who assert their right to belong and participate in public life as equals. Another positive account of the politics emergent from the exclusion of the urban poor is Robins' (2014) account of the 'slow activism' of rights-based social movements in South Africa in contrast to insurgent 'politics of spectacle', which burn bright for a moment but often have limited policy impact. Based on an alliance across class lines, 'slow activism' combines the tactics of the 'spectacular' with substantive policy engagement through democratic channels, indicating a long-term strategy for social change.

While Holston and Robins offer relatively optimistic accounts of the politics possible for the urban poor in the Global South in the name of democratic equality, most commentators are more pessimistic. A famous account is Partha Chatterjee's (2004, 2011) contrast between 'political society' and 'civil society', originating in India. Chatterjee argues that conditions for democratic politics from the poor in 'political society' are undermined in two ways. First, following Foucault, he holds that bureaucracy regards the urban poor as 'biopolitical' objects of development in ways that render them as 'populations to be governed' rather than as citizens bearing rights. Therefore, some organisations present themselves to the state in terms of populations of the needy rather than entities fighting for rights to secure key resources. Second, Chatterjee observes that many members of political society cannot live fully by the rules of the city, and this illegality makes them vulnerable to legal prosecution and police harassment and so less likely to engage the state democratically.

Chatterjee's insights alert us to forms of 'subaltern' or poor people's politics that operate more in terms of the logics of developmental governance of groups rather than democratic rights of citizens, and resolve into some form of clientelism or patronage mediated by political parties. As we have argued in respect of South Africa (Piper 2015; Piper & Anciano 2015; Anciano 2017), the mediation of the governed is enabled by a dominant party that seeks to control not just the state, but also the key forms of social representation to the state. In this context, popular protest may sometimes be a product of mediated politics, not just insurgency, such as when elite networks compete for control over key positions to mediate access to state resources (Piper & von Lieres 2015). Thus protest in and of itself is not necessarily a sign of political empowerment or a desire to disrupt in the name of equality.

Perhaps the most pessimistic account of the exceptionalism of the politics of the urban poor is found in Asef Bayat's work. Writing of non-democratic contexts in cities of the Middle East, Bayat (2000, 2013) develops the notion of 'quiet encroachment', where poor migrants settle in areas illegally or pursue livelihood practices through stealth, trying to avoid the gaze of the state altogether. If enough are able to settle in a place before the state notices, they may be able practically to defend their right to live in that place. It is only at the moment where encroachers are confronted by the state that they might organise collectively to resist. This then is the first moment of conscious politics in the practice of 'quiet encroachment'. While Bayat's account is drawn from undemocratic cities in the Middle East, his views have wider resonance, as many of the urban poor cannot afford to live by the rules of the democratic city. This exclusion through poverty means that many live in what Yiftachel (2009) terms the 'grey space' between legality and illegality. For these groups, participation in mainstream politics from a position of legal, social, and political vulnerability is a difficult option.

The constriction and capture of democratic spaces by elites

While the exclusion of important social groups from politics, most notably the urban poor, has clear implications for the democracy component of the liberal democratic model, there are two important additional challenges that face democratic institutions in the urban South. The first is the constriction of democratic institutions from elections through to participatory spaces by authoritarian rulers and social groups. Democracy is constricted when, for example, elections are cancelled or postponed, or some political parties or leaders are banned from participating. The significance of the constriction of the liberal democratic model is confirmed by a quick review of the regimes across the Global South. This reveals that around one-third are authoritarian, mostly the poorer countries of Africa and Asia such as Algeria, Burundi, Democratic Republic of Congo, Ethiopia, Laos, and Vietnam. Another one-third are identified as hybrid or illiberal regimes (Bolivia, Bhutan, Benin, Morocco, Thailand, Turkey), with only the final third being democratic in some form (Argentina, Botswana, Brazil, India, Indonesia, Malaysia, Mongolia, South Africa) (Democracy Index 2017).

If accurate, this means that in approximately two-thirds of all countries in the Global South there are significant constrictions of democratic institutions in that elections are not free, fair, or frequent; there are no opposition parties or they are harassed; there is a weak judiciary; and there is limited to no press freedom. This pattern is echoed in the list of the top twenty mega-cities in the world, where around half also have hybrid or authoritarian governments including Beijing, Cairo, Karachi, Moscow, and Shanghai (Democracy Index 2017; Demographia 2017). Again, we can assume constriction of elections, civil and political liberties, freedom of the press, and the capacity of the bureaucracy to implement policies. Lastly, as one would expect, although with some exceptions, in the Global South the majority of authoritarian and hybrid regimes have little participatory democratic innovation in their cities. Rather, the vast bulk of democratic innovation with participatory democracy is found in the middle-income democracies of Argentina, Brazil, India, Mexico, and South Africa.[1]

The second concern with democracy is the capture of democratic institutions, especially participatory institutions, by powerful social groups, whether political parties, corrupt officials and their wealthy sponsors, patrons and clients, or middle-class networks. Important to understanding why this matters is that much of the impetus behind the introduction of participatory democracy in the Global South is to ensure better influence of citizens over decisions that affect them directly, especially on grounds of poverty and marginalisation (Pateman 1985; Held 2006). This is especially the case in the Global South given more recent histories of 'differentiated citizenship' and the struggle for the 'right to have rights' (Dagnino 2003; Holston 2008:).

However, despite this intention, significant evidence is emerging of how new participatory institutions are captured by local political, economic or social elites (Piper & Deacon 2007; Kundu 2011; Teeffelen & Baud 2011; Patel, Sliuzas, & Georgiadou 2016). In addition, recent literature suggests that some middle-class groups have effectively supplanted poor and working-class groups in participatory spaces in India and Brazil (Chakrabati 2007; Avritzer 2017). Divided from poorer groups over issues such as security, the environment, and aesthetics, wealthy and middle-class groups use these spaces to advance their particular vision on these issues. In addition, these groups often also have the resources to combine this capture with forms of 'legal democracy' favoured by the right (Held 2006: 264). (This is where rights are defended through the courts, which tends to be an expensive and slow process.) The capture of new participatory spaces in these ways undermines their capacity to deepen democracy by better including the traditionally under-represented urban poor in local rule.

The limited power of City Hall over local governance and informal life

The third set of challenges for democratic rule in the Global South concerns the capacity of elected officials to take the important decisions that affect the city. While City Hall is nominally the highest authority over the people and places that constitute the urban, this power is often limited in the Global South, both by new forms of 'co-governance' and by enduring informality. As we have already suggested, the question of the power of City Hall is the key challenge that faces both inclusive and effective governance, and impactful democracy in the urban South.

A key trend in global politics post-Cold War is the decentralisation of various forms of authority from the central state to local levels, as part of the democracy export business. Indeed, decentralisation is often linked to democratisation, from the introduction of multi-party politics at the local level through to the innovations around participatory democracy described above (World Bank 2004). In addition to these forms of political decentralisation, other forms of decentralisation also exist, namely administrative, fiscal, and market. According to the World Bank (n.d.):

> all of these forms of decentralization can play important roles in broadening participation in political, economic and social activities in developing countries. Where it works effectively, decentralization helps alleviate the bottlenecks in decision making that are often caused by central government planning and control of important economic and social activities.

Most arguments for decentralisation are grounded on ideas of building a more inclusive, capacitated, and efficient system of governance at the local level. Notably, a key part of this framing is the option of shifting traditional government functions to quasi-independent organisations or the private sector. Indeed, for most observers, decentralisation is part of a wider package of policy measures, alongside privatisation and deregulation, that characterises the dominant framework for governance in the Global South after the Cold War. Considered together, these policies arguably create conditions for the decentring of political power at the local level from City Hall, facilitating the rise of 'governance' rather than 'government' as multiple actors collaborate to co-ordinate rule, giving rise to notions of 'co-governance', as well as 'network' and 'nodal' governance (Blanco 2015: 123).

The emergence of these forms of co-governance is seen as weakening the relative power of City Hall over local rule for three reasons. The first is that co-governance is an ideology that conceals the neo-liberal rule of the urban in the interests of business (Offe 2009). As Blanco (2015: 124) points out, critics of the governance paradigm hold that 'rather than the development of new plural, horizontal and inclusive forms of network governance, critics say, what we observe in European cities is the increasing concentration of urban power in the hands of a few political and business elites'. A similar set of arguments proliferate across the Global South including accounts of neo-liberal governance of solid waste collection in Cape Town (Miraftab 2004), garment production in Mumbai (Mezzadri 2008), education policy in Latin America (Torres 2002), and housing in Istanbul (Lovering & Türkmen 2011) and Jakarta (Yunianto 2014). Key to most accounts is the asymmetry of power between market and other social actors that means co-governance tends to be market-friendly governance. This theme echoes Stone's (1989) account of the resource advantage of business in gathering the power required to produce the city.

The second reason co-governance weakens the relative power of City Hall is that new forms of co-governance may work according to logics antithetical to democracy (Brown 2015; Swyngedouw 2007). Central to this idea is that forms of co-governance could have an ontology of the social, a mentality and a set of objectives and rules that run against democratic principles. Thus, neo-liberal or market governance might imagine the distribution of housing as an issue of individual choice, formalised through contract law upheld by the state, and operationalised through the metric of economic exchange for profit. On this view there is limited to no space for the idea of housing as a collective good, meeting an enduring and contextual human need, and subject to collective decision-making through a political process (see Brown 2015). Conversely, the co-governance of housing framed in developmental terms might imagine the problem as one of the distribution of shelter to maximise the well-being of a needy population (see Chatterjee 2004). On this view, residents are not citizens bearing rights but a group without the capacity to govern itself productively, and thus in need of external patronage. The general point is that different forms of co-governance might have different logics, not all of which are consistent with democratic principles.

The third reason for concern is the potential for erosion of the democratic relationship between elected officials and the citizens who elect them by new partners in co-governance. This issue includes the corruption of government officials by particular businesspeople, but also the transformation of governance in more systematic ways as expressed in theories of patronage, clientelism, prebendalism, and neopatrimonialism. These accounts centre on a 'gap' between the formal rules of governance and the actual practices of government officials, which is better explained in terms of theories of reciprocal political exchange under some conditions of dependency (Lemarchand & Legg 1972; Scott 1977). The problem from a democratic point of view is that these forms of rule reduce the accountability of elected officials to voters. This is because the distribution of goods may not happen through formal institutions but rather through personalised networks in exchange for support from key groups (Daloz & Chabal 1999).

In addition to the limitations on the power of elected officials posed by the rise of forms of co-governance, City Hall in the urban South must also confront the problem of large numbers of urban residents, mostly the very poor, living partially outside the rules of the formal system. As noted by Ledeneva (2018: 343), informal political relations are characterised by both techniques of co-option and control. Co-option includes special access to state resources through informal means such as corruption or patronage, often expressed in popular vernacular in terms associated with food. For example, *kormlenie* (Russia) literally means feeding, *kula* (Tanzania) means eating, and *uhljeb* (Croatia) means bread. Simply put, you pay or play along and you 'eat'. Conversely, control is exercised through what Gel'man terms 'the politics of fear', where selective repression is used to dissuade residents from resistance, often by making an example of a key figure. Examples include 'politically driven arrests ... exile ... torture, the disappearance of people and political assassinations' (Gel'man in Ledeneva 2018: 420).

Furthermore, informal power is not just about informal influence over formal institutions, but also about forms of power that exist outside the formal realm, in the grey zone between formal system and private life. It is possible for both informal actors, such as gang members, to make key decisions about the distribution of key resources on an ad hoc basis, and for informal systems of rules to be administered by these informal actors too (Wheeler 2014). Thus, Arjona (2014) argues that even in war zones rebel groups may set up informal institutions, understood as sets of publicly known rules, to distribute key social goods. Indeed, her study of dozens of cases in Colombia revealed that rebel groups often establish informal institutions that cover issues beyond security and taxation, including the local economy, social relations, and private conduct. These she terms *rebelocracy* and contracts with *aliocracy*, where armed groups seek to monopolise violence and tax residents to support this.

Lastly, it may even be possible to identify institutions, or more accurately sets of rules (North 1990), that are informal but not enforced by any clear ruler, such as when social norms guide forms of behaviour in ways contrary to the rule of the formal system as expressed in Simone's (2004) idea of 'people as infrastructure'. Cases like this would arguably also include the popular enforcement of conservative norms around sexuality such as 'corrective rape' (Bartle 2000). More compelling examples of informal rules without rulers are found in Ostrom's (1990) *Governing the Commons*, where public resources are managed collectively through rules drawn up, not by the state, but by local voluntary organisations, and which are sometimes informal but observed. Notably, such examples of 'rules without rulers' are rare.

In summary then, a central theme in the literature on governance in the urban South is the limited power of City Hall over the urban, both due to the tendency to share power with business and civil society, but also due to the emergent forms of informality of the urban poor, made more common by rapid urbanisation.

Theorising a framework for analysing democratic rule in the urban South

We began our argument by identifying the paradox of the emergence of participatory institutions, mostly but not exclusively in the Global South, at the same time as satisfaction with rule by democratic government is declining. Noting the mainstream model of liberal democratic urban rule as one of a virtuous feedback circle between politics, democracy, and local government, we identified three key empirical challenges to this model across the urban South. There are the exclusions from city politics, especially of the urban poor, the constriction and capture of democratic institutions, and the limits on the power of City Hall posed by both the rise of new forms of co-governance and enduring forms of informality. On this basis, we can refine our general problem into the following research questions:

- How democratic is local rule in the Global South, conceptualised in terms of a liberal democratic model of a virtuous circle of politics, democracy, and government?
- More specifically, in light of the current literature:
  - Is city politics inclusive of all major groups?
  - Are democratic institutions operational and open to all?
  - Do elected officials have the power necessary to govern?

To further develop an analytical framework for our study, we need to operationalise the key concepts in these questions in light of appropriate theory. To this end, we engage two distinct theories of power, tempered with insights from Foucault. The first is power as domination, drawn from Steven Lukes (2005), and the second is urban power as social production, from Clarence

Stone (1989). Lukes helps operationalise both political marginalisation in the urban South, and elite closure and capture of democratic institutions in terms of repression, concealment and depoliticisation. In turn, Stone helps frame the power of city officials in terms of the capacity to or 'power to' create the city under conditions of rapid social change, especially urbanisation.

Lukes and domination

To help conceptualise relations of domination we use Lukes' (2005) account of power that identifies three dimensions or 'faces' of power. The first dimension of power involves overt repression when one individual or group can impose their will on another (A has power over B when A can get B to do something B would not otherwise do) (Lukes 2005: 16). The focus is on 'behaviour in the making of decisions on issues over which there is an observable conflict of (subjective) interests, seen as express policy preferences, revealed by political participation' (Lukes 2005: 19).

The second dimension of power involves agenda-setting power, where a group can manipulate the political process so that their rivals are unaware of decisions being made that would harm them, or a powerful group can prevent certain issues becoming public knowledge (A has power over B when A controls the agenda of decision-making to A's advantage/B's disadvantage). In contrast to the one-dimensional view of power, this account sees power as exercised not only through decision-making but precisely through the act of 'non-decision making' too (Lukes 2005: 22). Thus, in addition to moments of overt conflict over issues in decision-making, this two-dimensional view also considers power to be exercised when 'decisions are prevented from being taken on potential issues over which there is an observable conflict of (subjective) interests, seen as embodied in express policy preferences and sub-political grievances' (Lukes 2005: 25).

The third dimension of power involves the establishment or maintenance of power relations by affecting preferences so that some groups do not understand that they are repressed (A has power over B when A affects B contrary to B's interests). In contrast to both preceding accounts of power, this view holds that power may be exercised not only through winning at decision-making, or controlling the agenda of decision-making, but also through preventing people from understanding that their interests are being harmed. Indeed, this transformation of consciousness might work so that neither the beneficiaries nor the victims of this form of power are fully aware of how their interests are advanced or harmed, respectively (Lukes 2005: 28).

Using Lukes' three-dimensional account of power as domination, we can operationalise the concept of exclusion from politics in the urban South in terms of repression, manipulation, and depoliticisation. These three concepts correlate to Lukes' three faces of power. Thus, domination as repression refers to instances where rulers or dominant groups overtly prevent groups who are trying to speak, organise, or mobilise from doing so. Examples could include banning political meetings, arresting opposition leaders, and refusing permission for political marches or peaceful protest. Domination as manipulation proceeds when the oppressed group is kept ignorant about decisions that affect them negatively, or is kept unaware of issues that harm them; thus, they do not politicise these issues. Examples could include when rulers conceal their interests from the public, fund rivals to critical civil society organisations, or secretly harass activists.

Lastly, domination as depoliticisation occurs when certain groups do not self-consciously participate in public debate, organise, or mobilise to advance their cause because they fail to perceive that they have common interests at all. A reason for this could be mechanisms of governance that, following Foucault (1980, 2000) embody particular forms of knowledge/power that shape subject positionality and subjectivity in depoliticising ways. Key here is the framing of the problem of power in knowledge terms. Thus, if the problem of power is the most efficient means to enable exchange, actors need to be framed as individual subjects such as consumers, rather than in the collective terms of the nation or the people. This framing of political subjectivity makes political organisation more challenging. Examples include when issues of public distribution are constructed as problems of individual choice, when organisations present as pro-poor but in reality are accountable to wealthy donors, and when outsiders are scapegoated through direct action for problems inherent to the local system of governance.

Similarly, we can operationalise elite control of democracy in the same terms of repression, manipulation, and depoliticisation. Repression refers to those instances where rulers or dominant groups prevent citizens from criticising rulers, or from organising opposition political parties, or fail to hold free, fair, and frequent elections. In all these instances there is overt conflict where the powerful dominate the will of opposing groups. Manipulation, on the other hand, typically avoids direct conflict by: concealing the true interests or actions of rulers from voters; secretly leaking damaging information on political rivals; deliberately failing to implement democratic reforms such as enabling participatory space properly; and cherry-picking decisions from democratic forms that suit the rulers and avoiding those that do not. Lastly, depoliticisation refers to political belief systems that serve to undermine potential opponents' claims to rule by, for example, portraying radical political parties or social movements as undemocratic and thus outside of normal political bounds. Used in this way, even the discourses of human rights and democracy can serve an ideological end (Hamilton 2003, 2014; Wood 2006), preventing marginalised groups from identifying with, joining or voting for radical movements and parties.

Stone and social production

Lukes' account of power is one that focuses on relations of domination. A key reason for this, as Foucault points out, is that until the twentieth century power was associated with a central sovereign state exercising control through law under the threat of punishment (Rose 1999). However, in recent times, alternative conceptions of governance beyond this notion of sovereignty emerged, not least through Foucault's account of disciplinary and regulatory/ security power (Collier 2009). In addition, these alternatives raise questions about the adequacy of a Lukesian conception of power as domination. In respect of urban politics, a parallel shift has occurred in thinking about the problem of power and governance as not simply one of sovereign control but more one of co-operative social production. Key here is the work of Clarence Stone (1989).

In *Regime Politics: Governing Atlanta* 1946–1989, Clarence Stone hypothesised that city power in America should be conceived in terms of a distinction between 'power over' decision-making and 'power to' implement decisions. This was because while cities had the authority to make local laws and policies, business had the resources to implement many key decisions through having the capital that generates jobs, tax revenues and project financing. Effective cities, therefore, were those that could form an informal alliance or 'regime' between political and economic elites to combine City Hall's formal control over law with businesses' resources. For Stone, the power required to rule the city is not so much one of control but more of social production. It is about developing the capacity to create the urban where it does not exist through an informal partnership between political and business elites.

While urban regime theory is regarded as specific to US cities in the 1980s and 1990s, the discourse of governance as co-operation between actors where the state 'steers rather than commands' (Stoker 1998: 17) is also found in urban governance theory advanced mostly by European scholars (Davies 2003; Le Gales 2002; Pierre 2011, 2014). Pierre points out that urban regime theory was not a popular theory in Europe as many cities there enjoyed increasing powers to tax versus the national state, unlike in the USA. The national policy framework was also stronger, reducing the tendency both for cities to rely on business for capital, and to compete with each other for business investment. Thus, many of the assumptions underlying urban regime theory in the USA simply do not apply in Europe.

Taking this into account, urban governance theory also moves from the assumption that the local government cannot rule entirely by itself, and thus partnership with other actors is required, resulting in 'governance' or 'co-governance'. Indeed, it is evident from the large urban population living under conditions of informality that urban rule in the Global South is facing a challenge of social production – literally the power to create the city. Urban governance theory is thus the idea that the city shares 'power with' other actors by combining 'power over' decision-making with 'power to' implement decisions in multiple ways. Notably, urban governance theory is more flex-ible than urban regime theory on who makes decisions and who implements them, and identifies four variables as key (civil society, national versus local authority, political economy, and globalisation) considerations in any analytical framework.

Similar to urban governance theory is the idea of network governance that posits an alternative 'horizontal' model across 'vertical' political and economic hierarchies. Thus, Newman (2005: 85) writes that 'the idea of a shift from markets and hierarchies towards networks and partnerships as modes of coordination is a dominant narrative'. This 'third way' narrative is embraced by a variety of authors who see networks as holding the potential to overcome 'the limitations of anarchic market exchange and top-down planning in an increasingly complex and global world' (Jessop 2003: 101–102). The idea is that network governance potentially extends the realm of public debate and engagement, empowering urban residents through inclusion in new relations of power, thus building new spaces for policymaking.

Another theoretical take that resonates with network theory is the 'nodal' theory of security governance (Shearing & Wood 2003; Wood & Shearing 2013). Noting the rise of privately owned but publicly used spaces, such as shopping malls and airports, and non-state provision of security in these spaces, nodal theorists hold that governance is better understood in terms of 'nodes'. In these nodes public and private actors network together to form 'institutions with a set of technologies, mentalities and resources – that mobilize the knowledge and capacity of members to manage the course of events' (Drahos, Shearing, & Burris 2005: 33). Again, this can be seen as a form of power as social production, but not one that necessarily directly involves the local state at all. Thus, all these accounts of urban power draw attention to the plurality of ways that City Hall might be involved in urban rule, from governing directly through partnering to deliver services, to creating the enabling environment for others to act.

The argument concerning urban power as social production through co-governance in some forms assumes that governance is well ordered into systems with rationalities, albeit in complicated, even contradictory, ways. While this approach might make sense for theorists of Northern modernity like Foucault, it makes too many assumptions about the urban South. Key here is the observation above of the limits of formal rule altogether, especially given the large populations of the urban poor in most cities around the world. The question of order arises in the context of the urban South separately from the question of authority, in other words, who governs, or whether they govern democratically or in an authoritarian way. Thus, in addition to the problem of power as social production in the urban South, the issue of power as social control endures through informal life (see Figure I.3).

To illustrate this claim more systematically, we note that a well-ordered form of governance is one with well-known rules, whereas disorder is the lack of a set of common rules akin to Hobbes' account of the state of nature. Indeed, some even argue that certain elites in the Global South have an interest in disorder to a degree, as it enables them to camouflage their self-aggrandisement, and maintain their status and power (Daloz & Chabal 1999). This granted,

	Forms of power		
1. Exclusion from politics?	*Repression*	*Manipulation*	*Depoliticisation*
a. Public debate	Is speech banned, access to meetings or public debate prevented?	Do political groups conceal their interests or activities from their rivals or the public?	Are subjectivities or interests framed in non-political ways?
b. Organisation	Are organisations banned, leaders arrested?	Do elites secretly funding rival groups, or establish 'sweetheart' organisations to divide base?	Do leaders present as pro-poor when really accountable to wealthy donors?
c. Mobilisation	Is permission for marches refused? Is violence threatened? Are peaceful protestors arrested?	Is violence threatened through surrogates? Are websites hacked? Are there anonymous death threats?	Are outsiders scapegoated for problems inherent to governance system?
2. Closed or captured democracy?	*Repression*	*Manipulation*	*Depoliticisation*
a. Elected office	Can residents criticise rulers? Can residents form new political parties? Are elections free, fair & frequent?	Do rulers conceal their interests of activities from the public? Do rulers secretly leak damaging information on their rivals?	Do rulers criticise rival contenders as undemocratic?
b. Influence officials	Are the legal requirements for participation implemented? Do rulers repress social movements? Do rulers threaten critical media outlets?	Is participation poorly implemented on purpose? Do rulers cherry-pick from participatory forums or civil society submissions?	Do rulers criticise radical social movements as undemocratic?
3. Empowered elected officials?	*Limited control*	*Dependency on social production*	
a. Events	To what degree do local officials decide on significant events (disaster response, land invasions, sport tournaments)	Who must co-operate, and how, to ensure the adequate implementation of key events?	
b. Rules	To what degree do local officials decide the rules for social practices by themselves, in partnership, or not at all?	Who co-operates, and how, to enable processes and rules for social practices?	

Figure I.3 Analytical framework for democratic local rule in the Global South (with sample questions)

it is important not to overstate the extent of disorder, for as Arjona (2014) notes, even in war situations some kinds of institution (conceived of as explicit rules) exist, often far more extensively than commonly assumed. Similarly, it is a mistake to assume that informal life in the urban has no rules, or that these rules are not related to the formal in some way.

Thus, one way of thinking through the relationship between the formal and informal is through the idea of meta-rules. These can be the foundational rules, even informal constitutions, that govern when decisions will follow formal or informal rules as Newton argues (in Ledeneva 2018). Thus, meta-rules will 'tell civil servants when they can relax, suspend or modify rules, in respect of which persons, and in which circumstances. Those meta-rules can also be considered part of the informal constitution' (Ledeneva 2018: 475). This example refers to instances like corruption, or patronage linked to formal office. But much of informal life in the Global South proceeds without the state, even though it may still be governed by a rule, such as one that designates who the non-state authority is in a certain context. An example: when poor migrants move into Cape Town and rent a place in the informal settlement in Imizamo Yethu they may be required to approach an informal local leader with a letter of character reference. Thus, even contexts with limited formal rules might have meta-rules or processes that define who governs.

This insight means that to the urban governance challenge of the social production of power that Stone notes, in the urban South we must add the problem of formal control. Simply put, where informality exists, no one regime of rulers or even system of governance can ever fully be in control of decisions on either key events or the rules of social practice. Governance in the urban South is always in some lesser or greater part a story of excess, as it always confronts the limits of governability.

How is local democratic rule experienced? Ask the residents of the city

By now it is clear that exploring democratic rule in the urban South is a multi-dimensional and complex undertaking that must involve probing politics, democratic institutions, and government, and the relationship between them. It is also clear that to understand the problem of a democratic deficit citizen's views must be a central focus of enquiry. Indeed, it is precisely the difference between the popular demand for democracy and the experience of democratic government that instigated our enquiry in the first place. Hence, to meet these multiple research objectives, we intend exploring the operation of democratic rule in one study site and from the citizens' point of view.

Exploring local democratic rule from the citizens' point of view implies a largely qualitative research design as the project explores complex relations between perceptions and experiences of politics, democracy, and governance. Perhaps an emergent theory of urban democratic rule can be operationalised for quantitative study in the future, but for this project, the evaluative and exploratory nature of the research problem required ethnographic case study research at the local level. Further, to accommodate the emergent effects of a system of rule it is important that the case study is large enough to include the potential diversity of forms of governance, as well as local democratic institutions. At the same time, it needs to be manageable in terms of the scale of research required.

To accommodate these ends we decided to explore the politics, democracy, and governance of a settlement, or group of settlements, in Hout Bay, part of the City of Cape Town. Historically a village in a valley adjacent to the city centre, Hout Bay now has a population of around 35,000 residents which makes it a manageable size, but it comprises a diversity of residents and residential areas, including a business and shopping district, that make it something of a microcosm of Cape Town. In racial, socio-economic and nationality terms, Hout Bay includes the majority of the key racial and social groups found more widely in the city, and indeed the country. In addition, by focusing on the suburban level we are able to investigate urbanisms more effectively, as it is at this micro-level that people live, work, and interact (or do not).

If the politics is inclusive and the institutions of local democracy work, this is the level at which we would expect them to have purchase. Indeed, Hout Bay's suburban boundaries correlate mostly with local government electoral ward boundaries; it has a lively civil society and political party branches with diverse groups living in close proximity, and significant local resources, all of which should enable local agency. Lastly, we focus on the suburb of Hout Bay for a significant period, from 2011 to 2017. This period covers several iterations of local and national government elections, and allows us to explore many key incidents, conflicts, and dynamics in sufficient detail to construct an account of the different forms of governance that are manifest at the micro-level, the relationship between them, and that with local democracy.

In terms of research design, we have already made the case for why a qualitative study using an exploratory case study method is appropriate. However, we need to say more about the strategy for answering the research questions. Given our focus on politics, democracy, and governance, we decided to begin each chapter with an instance of political conflict as a way into the exploration of democracy and governance in Hout Bay and the relationship between them. Although this follows the logic of our framing, it is also a choice made in homage to Robert Dahl's (1961) *Who Governs?* In this iconic text, Dahl attempts to repel Wright Mill's theory of the power elite by examining, in behavioural terms, conflict over formal decision-making in New Haven, Connecticut. Referencing Dahl's starting point of a power analysis as observable conflict, we thought we would start by identifying the main points of public conflict in Hout Bay, and then work backwards historically to work out what they tell us about politics, democracy, and governance.

Studying state–society relations from the perspective of public conflict invokes another tradition, the 'contentious politics' of social movements made famous by Charles Tilly (1986, 1993, 2006, 2008). Noting that democ-ratisation creates the conditions for contentious politics through the civil and political rights of freedom of speech, assembly, association, and protest, Tilly locates contentious politics in the context of the local evolution of popular politics linked to democracy. Informed by Lukes (2005) and Rossi (2017), we note the importance of analysing local politics beyond moments of public contention. Thus, while we start with these we trace the history of conflicts in Hout Bay through both contentious and constructive, as well as public, semi-public, and private moments. Consequently, we start with instances of observ-able conflict, yet we also strive to surface power relations beyond observable conflict insofar as they shed light on politics, democracy, and governance.

Furthermore, in exploring the politics around issues in Hout Bay, we look to unpack the democratic institutions at play, exploring instances of repression, manipulation, and depoliticisation, and the forms of governance that shape political engagement. In respect of the latter, we start with the actual rules of conduct around an issue, working to identify how governance happens, and by whom. This approach follows both Ostrom in defining institutions in terms of rules, and Foucault in placing practice and patterns of behaviour at the heart of governance, opening space for more various and nuanced accounts of who governs, but just as importantly how they govern. This sheds light not just on the role of elected officials in the urban, but also on the various sub-stantive ways in which the urban is ruled.

Reading *Democracy Disconnected*

The chapters of the book are designed according to the key issues that have dominated local politics in Hout Bay over the many years of our fieldwork, but also so as to offer keen insights into the large range of issues that frame the right to the city in this context. Issues such as water, human waste, and the environment; property, housing, and urbanisation; transport, and private development; education and public development; fishing, employment, and livelihoods; security, criminality, drugs; and the media, race, and nationality all feature in the book.

We suggest that you read Chapter 1 to get a clearer background on Hout Bay, its history, its people, and key features of local democracy and govern-ment. We also suggest referring to the conclusion for the integration of the findings of each chapter into the most sophisticated version of the 'dem-ocracy disconnected' argument. Then dip into the chapters that grab your attention. We have taken care to bring to life the politics around each issue in an engaging way, as well as using these to shine a light on the disconnect between democracy and governance in Hout Bay.

We have also taken care to ensure an even spread of chapters across the people and places that constitute Hout Bay, from the densely populated, poor, and black settlement of Imizamo Yethu with its diverse nationalities, vibrant street life, significant informality and strong identification with the African

National Congress political party; through the spacious, wealthy, white 'Valley', with its security estates, malls, 'European swallows' and residents associations; to the picturesque but depressed 'coloured' settlement of Hangberg, with its fishing community, smuggling, and housing concerns.

Hout Bay is a fascinating and intriguing place that hosts many characters good, bad and in between. We have changed names and protected the identities of respondents and local leaders as best we can, unless explicit consent to be named was requested. We do this not just for ethical reasons but also because our objective is to reflect on the implications of politics in Hout Bay for understanding urban politics more widely. Thus, while this book would be impossible without the people and places of Hout Bay, the larger analysis about a topological framework of multiple and contending forms of urban governance (FUG) and its implications for democracy in Cape Town more widely are applicable in every major city in South Africa. Indeed, we would suggest that the story of democracy disconnected is, at least to some extent, the story of every city in the Global South.

Note

1   For examples, see https://participedia.net/.

References

Almond, G., & Verba, S. (1989). *The Civic Culture: Political Attitudes And Democracy In Five Nations*. Princeton, NJ: Princeton University Press.

Anciano, F. (2017). Clientelism as Civil Society? Unpacking the Relationship between Clientelism and Democracy at the Local Level in South Africa. *Journal of Asian and African Studies*, https://doi.org/10.1177/0021909617709487.

Arjona, A. (2014). Wartime Institutions: A Research Agenda. *Journal of Conflict Resolution, 58*(8), 1360–1389.

Avritzer, L. (2017). Participation in Democratic Brazil: From Popular Hegemony and Innovation to Middle-Class Protest. *Opin. Publica, 23*(1), 43–59.

Baiocchi, G., Heller, P., & Silva, M. K. (2011). *Bootstrapping Democracy: Transforming Local Governance and Civil Society in Brazil*. Stanford, CA: Stanford University Press.

Bartle, E. E. (2000). Lesbians and Hate Crimes. *Journal of Poverty, 4*(4), 23–43.

Bayat, A. (2000). From 'Dangerous Classes' to 'Quiet Rebels': Politics of the Urban Subaltern in the Global South. *International Sociology, 15*(3), 533–557.

Bayat, A. (2013). The Quiet Encroachment of the Ordinary. *Chimurenga*, 8–15. http://chimurengachronic.co.za/quiet-encroachment-of-the-ordinary-2/. Accessed 27 June 2016.

Benjamin, S. (2008). Occupancy Urbanism: Radicalizing Politics and Economy Beyond Policy and Programs. *International Journal of Urban and Regional Research, 32*(3), 719–729.

Blanco, I. (2015). Between Democratic Network Governance and Neoliberalism: A Regime-Theoretical Analysis of Collaboration in Barcelona. *Cities, 44*, 123–130.

Brown, W. (2015). *Undoing the Demos: Neoliberalism's Stealth Revolution*. Cambridge, MA: MIT Press.

Chakrabarti, P. (2007). Inclusion or Exclusion? Emerging Effects of Middle-Class Citizen Participation on Delhi's Urban Poor. *IDS Bulletin, 38*(6), 96–104.

Chambers, R. (1992). Rural Appraisal: Rapid, Relaxed and Participatory. *IDS Discussion Paper 311.* www.ids.ac.uk/files/Dp311.pdf. Accessed 10 October 2016.

Chatterjee, P. (2004). *The Politics of the Governed: Reflections on Popular Politics in Most of the World.* New York: Columbia University Press.

Chatterjee, P. (2011). *Lineages of Political Society: Studies in Postcolonial Democracy.* New York: Columbia University Press.

Coelho, V. S. (2006). Democratization of Brazilian Health Councils: The Paradox of Bringing the Other Side into the Tent. *International Journal of Urban and Regional Research, 30*(3), 1468–2427.

Cohen, J. (1997). Deliberation and Democratic Legitimacy. In J. Bohman and W. Rehg (Eds.), *Deliberative Democracy.* Cambridge, MA: MIT Press, 67–92.

Collier, S. J. (2009). Topologies of Power: Foucault's Analysis of Political Government Beyond 'Governmentality'. *Theory, Culture & Society, 26*(6), 78–108.

Cornwall, A., & Coelho, V. S. (2007). *Spaces for Change? The Politics of Citizen Participation in New Democratic Arenas* (Vol. 4). London: Zed Books.

Dagnino, E. (2003). Citizenship in Latin America: An Introduction. *Latin American Perspectives, 30*(2), 3–17.

Dahl, R. A. (1961). *Who Governs? Democracy and Power in an American City.* New Haven, CT: Yale University Press.

Dahl, R. A. (1973). *Polyarchy: Participation and Opposition.* New Haven, CT: Yale University Press.

Dahl, R. A. (1989). *Democracy and its Critics.* New Haven, CT: Yale University Press.

Daloz, J. P., & Chabal, P. (1999). *Africa Works: Disorder as Political Instrument.* Oxford: James Currey.

Dalton, R. J., & Wattenberg, M. P. (Eds.) (2000). Unthinkable Democracy: Political Change in Advanced Industrial Democracies. In *Parties without Partisans. Political Change in Advanced Industrial Democracies.* Oxford: Oxford University Press, 3–18.

Davies, J. S. (2003). Partnerships Versus Regimes: Explaining Why Regime Theory Cannot Explain Urban Coalitions in the UK. *Journal of Urban Affairs, 25,* 253–269.

Davis, M. (2006). *Planet of Slums.* London: Verso.

Democracy Index. (2017). *Democracy Index 2017. Free Speech Under Attack.* The Economist Intelligence Unit. https://goo.gl/1QYj1e. Accessed 3 March 2018.

Demographia. (2017). *Demographia World Urban Areas. 13th Annual Edition: 2017/04.* www.demographia.com/db-worldua.pdf. Accessed 3 March 2018.

Drahos, P., Shearing, C. D., & Burris, S. (2005). Nodal Governance. *Australian Journal of Legal Philosophy, 30,* 30–58.

Foucault, M. (1980). *Power/Knowledge. Selected Interviews and Other Writings, 1972–77.* Brighton: Harvester.

Foucault, M. (2000). The Ethics of the Concern for Self as a Practice of Freedom. In Rabinow, P. (Ed.), *Essential Works of Foucault 1954–1984, Vol. I: Ethics.* London: Penguin.

Frandsen, A. G. (2002). Size and Electoral Participation in Local Elections. *Environment and Planning C: Politics and Space, 20*(6), 853–869.

Gaventa, J. (2006). Triumph, Deficit or Contestation: Deepening the Deepening Democracy Debate. *IDS Working Paper 264.* Brighton, CDRC, IDS. www.ntd.co.uk/idsbookshop/details.asp?id=944. Accessed 18 December 2013.

Gaventa, J., & Barrett, G. (2010). So What Difference Does it Make? Mapping the Outcomes of Citizen Engagement. *IDS Working Paper 347*. Brighton: Institute of Development Studies. www.ids.ac.uk/files/dmfile/Wp347.pdf. Accessed 12 September 2016.

Gonçalves, S. (2013). The Effects of Participatory Budgeting on Municipal Expenditures and Infant Mortality in Brazil. *World Development, 53*, 94–110.

Hajnal, Z. L., & Lewis, P. G. (2003). Municipal Institutions and Voter Turnout in Local Elections. *Urban Affairs Review, 38*(5), 645–668.

Hamilton, A., James, M., & Jay, J. (1961). *The Federalist*, Edited by Cooke, J. E. Middletown, CT: Wesleyan University Press.

Hamilton, L. (2003). *The Political Philosophy of Needs*. Cambridge: Cambridge University Press.

Hamilton, L. (2014). *Freedom is Power: Liberty Through Political Representation*. Cambridge: Cambridge University Press.

Harvey, D. (2012). *Rebel Cities: From the Right to the City to the Urban Revolution*. London: Verso Books.

Held, D. (2006). *Models of Democracy*. Cambridge: Polity Press.

Holston, J. (2008). *Insurgent Citizenship: Disjunctions of Democracy and Modernity in Brazil*. Princeton, NJ: Princeton University Press.

Hughes, S. (2017). The Politics of Urban Climate Change Policy: Toward a Research Agenda. *Urban Affairs Review, 53*(2), 362–380.

Hunger Project & UN Democracy Fund. (2014). 2014 State of Participatory Democracy Report. THP. www.thp.org/wp.../2014/10/2014-State-of-Local-Dem-Report-The-Hunger-Project.pdf. Accessed 22 October 2016.

Jessop, B. (2003). Governance and Metagovernance: On Reflexivity, Requisite Variety and Requisite Irony. In H. P. Bang (Ed.), *Governance as Social and Political Communication*. Manchester: Manchester University Press, 101–116.

Johnson, C. (2015). Local Civic Participation and Democratic Legitimacy: Evidence from England and Wales. *Political Studies, 63*, 765–792.

Kundu, D. (2011). Elite Capture in Participatory Urban Governance. *Economic & Political Weekly, 46*(10), 23–25.

Le Galès, P. (2002). *European Cities: Social Conflicts and Governance*. Oxford: Oxford University Press.

Ledeneva, A. (Ed.) (2018). *The Global Encyclopaedia of Informality, Volume 2*. London: UCL Press. www.ucl.ac.uk/ucl-press/browse-books/global-encyclopaedia-of-informality-ii. Accessed 3 March 2018.

Lemarchand, R., & Legg, K. (1972). Political Clientelism and Development: A Preliminary Analysis. *Comparative Politics, 4*(2), 149–178.

Levitsky, S., & Way, L. (2002). The Rise of Competitive Authoritarianism. *Journal of Democracy, 13*(2), 51–65.

Lovering, J., & Türkmen, H. (2011). Bulldozer Neo-liberalism in Istanbul: The State-led Construction of Property Markets, and the Displacement of the Urban Poor. *International Planning Studies, 16*(1), 73–96.

Lukes, S. (2005). *Power. A Radical View*. Second Edition. Basingstoke and New York: Palgrave Macmillan.

Mezzadri, A. (2008). How Globalised Production Exploits Informal-Sector workers: Investigating the Indian Garment Sector. *Development Viewpoint*, (12).

Mill, J. S. (1859). *On Liberty*. London: Parker & Son.

Miraftab, F. (2004). Making Neo-liberal Governance: The Disempowering Work of Empowerment. *International Planning Studies, 9*(4), 239–259.

Morlan, R. L. (1984). Municipal vs. National Election Voter Turnout: Europe and the United States. *Political Science Quarterly, 99*(3), 457–470.

Newman, J. (2005). Participative Governance and the Remaking of the Public Sphere. In J. Newman (Ed.), *Remaking Governance: Policy, Politics and the Public Sphere*. Bristol: The Policy Press, 119–138.

Norris, P. (2011). *Democratic Deficit: Critical Citizens Revisited*. Cambridge: Cambridge University Press.

North D. (1990). *Institutions, Institutional Change and Economic Performance*. Cambridge: Cambridge University Press.

Offe, C. (2009). Governance: An 'Empty Signifier'? *Constellations, 16*(4), 550–562.

Ortiz, I., Burke, S., Berrada, M., & Cortés, H. (2013). World Protests 2006–2013. *Working Paper 2013*. Institute for Policy Dialogue and Friedrich-Ebert-Stiftung. http:// policydialogue.org/publications/working_papers/world_protests_2006-2013/. Accessed 22 October 2016.

Ostrom, E. (1990). *Governing the Commons: The Evolution of Institutions for Collective Action*. Cambridge: Cambridge University Press.

Parikh, J. K., Parikh, K. S., Subir Gokarn, J. P. Painuly, B. S., & Vibhooti Shukla, V. (1991). Consumption Patterns: The Driving Force of Environmental Stress. *Report prepared for United Nations Conference on Environment and Development (UNCED)*. IGIDR-PP-014, Mumbai.

Patel, S., Sliuzas, R., & Georgiadou, Y. (2016). Participatory Local Governance in Asian Cities: Invited, Closed or Claimed Spaces for Urban Poor? *Environment and Urbanization Asia, 7*(1), 1–21.

Pateman, C. (1985) *The Problem of Political Obligation: A Critique of Liberal Theory*. Cambridge: Polity.

Pierre, J. (Ed.) (1995). *Bureaucracy in the Modern State*. London: Edward Elgar.

Pierre, J. (2011). *The Politics of Urban Governance*. Basingstoke: Palgrave.

Pierre, J. (2014). Can Urban Regimes Travel in Time and Space? Urban Regime Theory, Urban Governance Theory, and Comparative Urban Politics. *Urban Affairs Review, 50*(6), 864–889.

Piper, L. (2015). From Party-State to Party-Society in South Africa: SANCO and the Informal Politics of Community Representation in Imizamo Yethu, Hout Bay, Cape Town. In Bénit-Gbaffou, C. (Ed.), *Popular Politics in South African Cities: Unpacking Community Participation*. Pretoria: HSRC Press, 21–41.

Piper, L., & Anciano, F. (2015). Party Over Outsiders, Centre over Branch: How ANC Dominance Works at the Community Level in South Africa. *Transformation: Criti cal Perspectives on Southern Africa, 87*(1), 72–94.

Piper, L., & Deacon. R. (2007). Too Dependent to Participate: Ward Committees and Local Democratisation in South Africa. *Local Government Studies, 35*(4), 415–433.

Piper, L., & von Lieres, B. (2015). Mediating Between State and Citizens: The Significance of the Informal Politics of Third party Representation in the Global South. *Citizenship Studies, 19*(6–7), 696–713.

Pogrebinschi, T., & Samuels, D. (2014). The Impact of Participatory Democracy: Evidence from Brazil's National Public Policy Conferences. *Comparative Politics, 46*(3), 313–332.

Putnam, R. D., Leonardi, R., & Nanetti, R. Y. (1994). *Making Democracy Work: Civic Traditions in Modern Italy*. Princeton, NJ: Princeton University Press.

Robins, S. (2014). Slow Activism in Fast Times: Reflections on the Politics of Media Spectacles after Apartheid. *Journal of Southern African Studies*, *40*(1), 91–110.

Rose, N. (1999). *Powers of Freedom: Reframing Political Thought*. Cambridge: Cambridge University Press.

Rossi, F. M. (2017). *The Poor's Struggle for Political Incorporation: The Piquetero Movement in Argentina*. Cambridge: Cambridge University Press.

Roy, A. (2011). Slumdog Cities: Rethinking Subaltern Urbanism. *International Journal of Urban and Regional Research*, *35*(2), 223–238.

Scott, J. C. (1977). *The Moral Economy of the Peasant: Rebellion and Subsistence in Southeast Asia*. New Haven, CT: Yale University Press.

Shearing, C., & Wood, J. (2003). Nodal Governance, Democracy, and the New 'Denizens'. *Journal of Law and Society*, *30*(3), 400–419.

Simone, A. (2004). People as Infrastructure: Intersecting Fragments in Johannesburg. *Public Culture*, *16*(3), 407–429.

Stoker, G. (1998). Governance as Theory: Five Propositions. *International Social Science Journal, 50*(155), 17–28.

Stone, C. (1989). *Regime Politics: Governing Atlanta, 1946–1988*. Lawrence, KS: University Press of Kansas.

Swyngedouw, E. (2007). The Post-Political City. In BAVO (Ed.), *Urban Politics Now: Re-imagining Democracy in the Neoliberal City*. Reflect Series. Rotterdam: Netherlands Architecture Institute (NAI) Publishers, 58–76.

Teeffelen, J. V., & Baud I. (2011). Exercising Citizenship: Invited and Negotiated Spaces in Grievance Redressal Systems in Hubli–Dharwad. *Environment and Urbanization Asia, 2*, 169–185.

Tilly, C. (1986). *The Contentious French*. Cambridge, MA: Harvard University Press.

Tilly. C. (1993). Contentious Repertoires in Great Britain, 1758–1834. *Social Science History, 17*(2), 253–280.

Tilly, C. (2006). *Regimes and Repertoires*. Chicago, IL: University of Chicago Press.

Tilly, C. (2008). *Contentious Performances*. Cambridge: Cambridge University Press.

Torres, C. A. (2002). The State, Privatisation and Educational Policy: A Critique of Neo-Liberalism in Latin America and Some Ethical and Political Implications. *Comparative Education, 38*(4), 365–385.

UN (United Nations, Department of Economic and Social Affairs, Population Division). (2014). *World Urbanization Prospects: The 2014 Revision, Highlights* (ST/ESA/SER.A/352). https://esa.un.org/unpd/wup/publications/files/wup2014-highlights.pdf. Accessed 3 March 2018.

UN Habitat. (2016*)*. *Slum Almanac 2015–2016: Tracking Improvement in the Lives of Slum Dwellers*. United Nations. https://unhabitat.org/wp-content/uploads/2016/02-old/Slum%20Almanac%202015-2016_EN.pdf. Accessed 3 March 2018.

Wampler, B. (2012). Participation, Representation, and Social Justice: Using Participatory Governance to Transform Representative Democracy. *Polity, 44*(4), 666–682.

Wheeler, J. (2014). 'Parallel Power' in Rio de Janeiro: Coercive Mediators Between Citizens and the State. In von Lieres, B., & Piper, L. (Eds.), *Mediated Citizenship: The Informal Politics of Speaking for Citizens in the Global South*. Basingstoke: Palgrave Macmillan, 72–92.

Wood, E. M. (2006). Democracy as Ideology of Empire. In Mooers, C. (Ed.), *The New Imperialists: Ideologies of Empire*. Oxford: Oneworld, 9–23.

Wood, J., & Shearing, C. (2013). *Imagining Security*. London: Routledge.

World Bank. (n.d.). 'What is Decentralization?' www.ciesin.org/decentralization/English/General/Different_forms.html. Accessed 8 May 2018.

World Bank. (2004). *World Development Report: Making Services Work for Poor People*. The International Bank for Reconstruction and Development. https://openknowledge.worldbank.org/bitstream/handle/10986/5986/WDR%202004%20-%20English.pdf?sequence=1. Accessed 3 October 2016.

Yiftachel, O. (2009). Critical Theory and 'Gray Space': Mobilization of the Colonized. *City, 13*(2–3), 246–263.

Yunianto, T. (2014). On the Verge of Displacement: Listening to Kampong Dwellers in the Emotional Economy of Contemporary Jakarta. *Thesis Eleven, 121*(1), 101–121.

1 The Republic of Hout Bay

A house divided

Hout Bay is a suburb about 20 kilometres south of the heart of Cape Town, located in a valley on the west coast of the Cape Peninsula. A natural bowl shape, Hout Bay is remarkably beautiful, framed to the west by the dramatic Sentinel Mountain peak, and flanked to the east by the Table Mountain range. The lower slopes of the valley are green and fertile, fed by the Disa River that runs into a natural harbour with an attractive swimming beach. The striking setting of Hout Bay, and its relatively close proximity to the city centre, make it a sought after residential and tourist location (see Figure 1.1).

Although there is evidence of inhabitation of the area prior to colonisation, there is no written history before Dutch settlement. The name Hout Bay is an Anglicisation of the Dutch *Houtbaai* (literally 'wood bay'), and stretches back to 1653. The area was so named for the quantity of excellent timber growing in its ravines. The fishing at Hout Bay extends back prior to colonialism; the first colonial fishing village was established in 1867 when a German immigrant, Jacob Trautmann, began to farm and fish in the area (SAHistoryOnline 2015). Until the 1950s, the land in Hout Bay was used primarily for agricultural purposes and the area was sparsely populated. In the 1950s, the lumber industry went into decline and fishing, including factories processing fish, became the largest industry in the area. Following the decline of the fishing industry in the 1980s, Hout Bay became an increasingly desirable residential area as Cape Town rapidly densified (SAHistoryOnline 2015).

Today, adjacent to the two public roads that lead down the valley to the heart of the settlement, are signs saying 'Welcome to the Republic of Hout Bay', (see Figure 1.2). Somewhat faded and dilapidated, the signs, like the idea of a self-governing community, have seen better days. According to the *Hout Bay Organised* website, the idea of the Republic was born in the late 1980s when local charities and business attempted a tongue-in-cheek effort at secession:

> The concept took root, with an official passport which was sold at roadblocks at the three entrances to Hout Bay. Some residents took the idea quite seriously and there was even a President, an anthem and costume. Several people were able to convince immigration officers around the world to stamp their Hout Bay passports.[1]

Figure 1.1 Aerial image of Hout Bay

Figure 1.2 Welcome to Hout Bay sign

Notably, the idea of a Republic emerged at a time when Hout Bay was a small community with a mix of coloured and wealthy white residents. It was a quiet suburb with low levels of political contestation. Today Hout Bay is a multi-racial, multi-national and socio-economically diverse suburb of some 35,000 people, more than half of whom are not white, and a sizeable minority of whom are not South African. Following 30 years of significant migration, Hout Bay has transformed from a quiet seaside village into a bustling suburban centre.

However, there is more to the idea of a Republic than a common identity and belonging; there is the notion of self-government too, as expressed symbolically in the Hout Bay 'passport'. While in the post-apartheid political order a collection of suburbs cannot be sovereign, there are good reasons to be optimistic about the capacity of residents of Hout Bay to influence decision-making over their area. Hout Bay residents have access to a formal democratic system with both representational and participatory mechanisms, as well as a system of civil liberties that enables civil society organisation, popular mobilisation, and public debate. Furthermore, and this makes it somewhat different from much of the rest of Cape Town, while most residents are poor, a high proportion are wealthy, and thus the settlement has significant local capacity to drive development for all. Consequently, there is much going on in Hout Bay. Many people are trying to do many different things, from international donors, to national and provincial government, to local civil society and residents. They initiate housing projects, build clinics, open schools, create markets, host festivals, build food gardens, conduct health education, clear alien vegetation, and collect rubbish. The list is endless.

Given this significant capacity, opportunity, and intent to make a real difference in public life, it seems safe to assume that local democracy in Hout Bay would be healthy too. Indeed, it is unlikely that there are too many suburbs in South Africa with conditions more conducive to effective and democratic local governance. However, as we will demonstrate, this is not case. Indeed, Hout Bay is close to an archetype for the democratic deficit described in the opening paragraph of the book. Thus, despite a dramatic increase in democratic institutions and opportunities since the end of apartheid, most residents of Hout Bay today feel frustrated by local governance. Indeed, when we started our research the one thing almost all respondents shared was a dislike of local politics, politicians, and government.

In this chapter, we will show how the idea of the Republic of Hout Bay is undone in two ways. First, despite economic, social and political relations across three adjacent communities, the area resembles the apartheid logic of spatially segregated communities divided by race, socio-economic status, and culture. Thus, the wealthy, white, and English-speaking suburb of 'the Valley' sits alongside the poor, black, isiXhosa-speaking settlement of

Imizamo Yethu, and both are close to the poor, coloured, and Afrikaans-speaking area of Hangberg (see Map 1.2). Second, despite the promise of local democracy, the Hout Bay community is not self-governing in any significant way; residents have limited influence over decision-making that affects their lives other than for the actions they take directly themselves, and the leaders of the three communities of Hout Bay often pursue contrary political ends.

The promise of a Republic

At one level, the idea that Hout Bay could be a self-governing Republic seems peculiar. The settlement is part of the City of Cape Town, which in turn is part of the Western Cape Province in the Republic of South Africa. The 1996 Constitution of post-Apartheid South Africa makes it clear in the very first founding provision that the 'Republic of South Africa is one, sovereign, democratic state' (RSA Constitution 1996). The numbers and boundaries of provincial and local government are decided by national legislation, and are not prescribed in the Constitution. There are currently nine provinces and 257 municipalities in South Africa.

While the Constitution divides political authority spatially in terms of three spheres of national, provincial, and local government, each with their own executive and legislative capacities, it privileges national over the other spheres, and gives limited powers to local government. Thus, Schedule 4 confers concurrent powers to national and provincial governments in respect of most key issues, including agriculture, health, housing, and policing. Section 44(2) gives national parliament primary power to frame any policy on any issue in the interests of national security, economic unity, setting national standards, or 'to prevent unreasonable action taken by a province which is prejudicial to the interests of another province or to the country as a whole' (RSA Constitution 1996). This affirmation of the primacy of the national sphere means, for example, that while in theory all spheres of government can raise tax, provinces do not currently do so. Furthermore, in respect of local government, Schedules 4b and 5b of the Constitution identify comparatively limited areas of competence for local governance, including street cleaning, local markets, air pollution, traffic, and the like.

While all this appears to concentrate power in the national sphere of the state, there are three sets of reasons why local municipalities, and especially metropolitan areas like Cape Town, possess a significant degree of power. The first is the Constitutional requirement, as set out in Chapter 3, for 'co-operative governance' that looks to reduce the possibility of gridlock or conflict by requiring constructive relations between the three spheres and the integration of rule across the state (RSA Constitution 1996). This requirement for co-operation is manifested in various attempts to integrate policy processes from local to national spheres, in particular around developmental governance (see section 41 of RSA Constitution 1996).

For example, the Municipal Systems Act of 2000 requires that each municipality develop a 5-year Integrated Development Plan (IDP), and that this plan incorporate the development priorities of provincial and national government (for instance around health clinics and schools) into one vision for the municipality (Local Government 2000). Further to the requirement of planning across spheres, the IDP is meant to address the socio-economic and spatial disparities that endure from apartheid. In respect of the latter, the recent Spatial Planning and Land Use Management Act (SPLUMA) of 2013 seeks to redress racial inequality, spatial segregation, and unsustainable land use patterns. Notably, Section 2(1)a of the requirements of SPLUMA apply to municipal planning, including Integrated Development Plans. Importantly then, municipalities are enabled by national legislation to drive both developmental and spatial planning processes.

In addition to this leading role around local development and spatial planning, municipalities are also empowered to raise taxes in the form of property rates and revenue through the sale of services. The City of Cape Town's total annual budget for 2017/18 was R44.3 billion ($3.25 billion) which is raised mostly from property rates (22 per cent), services such as electricity, water, sanitation, and refuse (52 per cent), and national subsidies (17 per cent) distributed in terms of an equitable grant system (CCT 2017). From this it is clear that, unlike most municipalities in the country (Ajam 2014), the City of Cape Town has a substantial degree of financial autonomy from national government as it raises the vast majority of its revenues locally. Key to this relative success is the fact that, unlike most of the rest of the country, Cape Town has both a long history of local government institutions, and a large population of over 3.75 million, a substantial proportion of whom can pay for both rates and services.

Last and probably least, Cape Town is governed not by the party that rules nationally, the African National Congress (ANC), but by the official opposition, the Democratic Alliance (DA). While the DA may claim to be better at governing than the ANC, in reality a moot point as illustrated by the recent threat of the City's taps running dry (Winter 2017), it positions itself publicly as better at clean and efficient governance than the ANC (Jolobe 2010). Given this, and the fact that until very recently, the DA had won just one of eight metros and one of nine provinces, the party is able to focus its resources on governing Cape Town in a way that the ANC simply cannot do in the many cities that it runs elsewhere in South Africa.

In short then, while the Constitution of post-apartheid South Africa is not based on the principle of subsidiarity, Cape Town nevertheless finds itself authorised to drive its own development and spatial planning. It also enjoys a significant degree of financial autonomy from the national treasury, in part due to its particular history, but mostly due to its scale and relative wealth. Lastly, it is governed by an opposition party that claims superior levels of clean and efficient government that can afford to invest its best people in running the city. Considered together, the legal, policy, and party-political conditions constitute a basis for practical control of Cape Town's future by City Hall.

The promise of democracy

For Hout Bay to approach the idea of a Republic, the processes of decision-making that govern daily life need to accommodate representation, participation, and planning in the name of Hout Bay. This 'long route' of accountability, to use the language of the World Bank (2003), is important because in Cape Town the 'short route of accountability', whereby residents engage directly with municipal officials, is not designed around spatial units such as Hout Bay. The City of Cape Town has over 40 line departments (for example City Health, Human Resources, Integrated Development Plan, Transport and Urban Development, Water and Sanitation). To make accessing services more efficient, the city has introduced an electronic C3 notification system where residents can report problems such as blocked drains, potholes, broken streetlights, water leaks, and so on, either online or with a phone call. In place since 2007, this system is generally regarded as a success (Bagui & Bytheway 2013). While laudable, a system like C3 does not allow Hout Bay residents to engage the city collectively. As a senior official in the waste sector explained:

> wards and subcouncils, those boundaries don't mean anything to us when it comes to our normal operations. We cannot honour invisible boundaries ... You can't say to a street sweeper here is a little boundary in the middle of this road, don't go past it ... Because the City is so big, we're divided into areas for management purposes ... the only time we take subcouncil or ward boundaries into consideration is when there are ward specific projects.
>
> (City official 1 2015)

Thus for collective input on substantive issues, residents must turn to the long route of accountability established through the system of local democracy in South Africa. This system has both representative and participatory components, as well as forms of civil society formation, popular mobilisation and various forums for public debate enabled by political liberties, all encoded in the 1996 Constitution (see Chapter 2: Bill of Rights).

In terms of democracy, the formal system of representation is a good place to start. Historically, Hout Bay had a degree of self-government during the transition from apartheid to democracy in the early 1990s. This structure was termed the Transitional Metropolitan Substructure of Llandudno/Hout Bay. By 2000, the institutions of local government had been finalised into a mixed electoral system where half the city councillors are elected from local districts called wards, and the other half through a party-list proportional representation system (section 157, RSA Constitution 1996). The proportional representation system on its own is used for provincial and national elections. Today the City of Cape Town (CCT) has 116 wards in total; one, ward 74, includes Hout Bay. In recent years, the boundaries of ward 74 have been redrawn more tightly around Hout Bay and Llandudno. Given that Hout Bay is far bigger than Llandudno, which has less than 1,000 residents, the ward councillor for ward 74 is practically the ward councillor for Hout Bay.

Importantly, developmental and spatial planning do not occur at ward level in the local government system in South Africa, but usually at a higher level. In smaller municipalities, the spatial level of planning will be the municipality as a whole, but in larger municipalities like Metropolitan councils, planning is also conducted at sub-municipal levels, which in Cape Town are called subcouncils. Hence the city is currently divided into 24 subcouncils constituting three to six adjacent wards. Hout Bay is part of subcouncil 16, which includes five wards along the Atlantic seaboard, as well as the City Bowl that is the symbolic heart of Cape Town (Map 1.1).

A Chairperson, who is a proportional representation (PR) councillor, and the political leader of the structure, heads the subcouncil. Subcouncils also include the ward councillors from the area and PR city councillors assigned to the area by their parties. The subcouncil Chairperson is elected by the members of the subcouncil in principle (CCT n.d.) but in practice this is a political appointment endorsed by the Mayor and his or her executive (Political representative 2 2017). The chairperson decides when and where the subcouncil meets, although they are meant to meet once a month, or 'whenever an urgent issues arises'. In practice, subcouncil chairs become 'like a

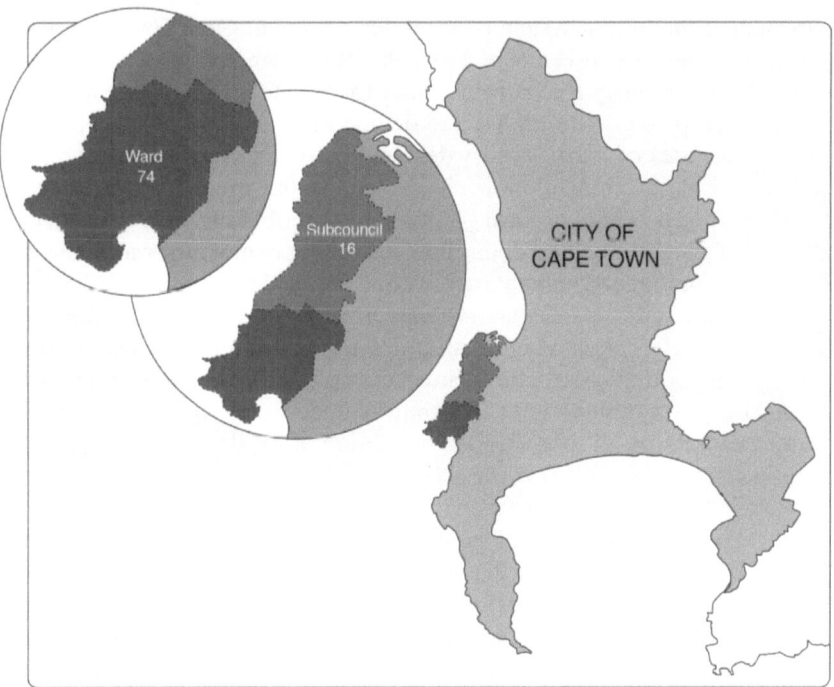

Map 1.1 Subcouncil 16, including Ward 74, in the City of Cape Town

mini-Mayor'. They have an executive council, a staff and an office (Political representative 2 2017).

In addition, each subcouncil has a manager appointed by the city administration who is responsible for co-ordinating the line departments behind designated subcouncil responsibilities. Importantly, subcouncils have authority over laws delegated to them by the City. These include most of the powers of local government, such as monitoring service delivery outcomes defined in the Integrated Development Plan, and budgeting. Until recently, subcouncils were also responsible for public participation processes but these became unwieldy because 'do you ask people's opinions on everything ... how long is a piece of string?' Thus public participation has now been moved to the four larger area-based political committees (Political representative 2 2017).

Thus, not only does subcouncil 16 have a central role in development planning for Hout Bay, it has a significant degree of jurisdiction over many of the issues that fuel local politics. As the lowest spatial level at which governance can occur across line departments, this is the closest decision-making forum through which attempts to advance Hout Bay-specific projects may take place. The difference in scale means that integrated governance planning and implementation at the level of Hout Bay can only occur within the larger frame of subcouncil 16. As illustrated in Chapter 2, this makes it difficult, but not impossible, to champion Hout Bay-specific development projects.

In addition to its representation aspects of the ward councillor and subcouncil, the formal system of local democracy in Cape Town offers substantial promise for residents of Hout Bay to influence city decision-making through participatory mechanisms. After the transition to democracy, South Africa attempted to decentralise and democratise local governance practices through the idea of 'participatory democracy'. The Deputy Minister responsible for local government reform in the late 2000s, Yunus Carrim, talked of the enormous potential of local government to mobilise people to participate actively in governance, and about how deepening democracy was a task that defined the 'national democratic transition' (Barichievy et al. 2005: 376).

This approach to participatory democracy is rooted in a long history of active civil society situated in urban centres, which challenged apartheid rule through representative structures. Street committees, civic associations, locally embedded trade unions and anti-apartheid networks such as the United Democratic Front, all subscribed to the notion of resident-based engagement in influencing governance (White 2008). The idea that popular participation is also intrinsically good resonates with the ideological position of several political parties in South Africa. The ANC, for example, prided itself as a party that elects leaders based on the wishes of its provincial, and in turn local, branch structures (Butler 2005).

There are several pieces of legislation that promote the implementation of participatory governance, including the Municipal Structures Act of 1998, and the Municipal Systems Act of 2000, which formalise in law requirements for 'participatory governance' that promise a public voice in governing between elections (Barichievy et al. 2005: 375). The ethos of these Acts is highlighted in the Department of Provincial and Local Government's (2007) *Draft National Policy Framework for Public Participation*, which states that public participation is an 'open, accountable process through which individuals and groups within selected communities can exchange views and influence decision-making'.

Two key participatory processes that fit within this legislated framework are public consultation on the Integrated Development Planning (IDP) process and the annual budget, and the representative structure of ward committees. The Municipal Systems Act requires that the municipality consult the public on the budget and IDP processes directly. For example, in 2016, the City of Cape Town conducted public meetings, addressed by the Mayor and Mayoral Committee in the eight spatial clusters of the city, at which the draft budget was discussed (PPU 2016). Unfortunately, public consultation on the budget and IDP processes are generally ineffectual in South Africa, with poor transparency and often low levels of participation (Njenga 2009; Penderis & Tapscott 2014). Key here is that public participation is limited to once-off public meetings on complex draft documents that cover the municipality or subcouncil as a whole. There is no space earlier in these processes of decision-making for meaningful public input, including by civil society, at a scale more relevant to residents. Consequently, public consultation on the budget and IDP is mostly an exercise governed by the need for official compliance.

Established in the Municipal Structures Act of 1998, ward committees are central to participatory democracy. Ward committees consist of ten individuals alongside the ward councillor, who is the chairperson of the committee. Each municipality will make the rules regulating the election or selection of members onto ward committees, although committee members should reflect a diversity of gender and interests. The statutory powers and functions of ward committees are very limited. They may only make recommendations on matters affecting the ward to the ward councillor or, through the councillor, to the council, executive committee, Mayor, or subcouncil. They can also identify and initiate local projects to improve the lives of people in the ward, monitor the performance of the municipality, and be involved in annual budget and IDP processes (DPLG/GTZ 2005). In principle, ward committees should increase the participation of local residents in municipal decision-making, represent the ward, and not be aligned to a party. In practice, research shows that they are better understood as extensions of ward councillor influence, and are usually partisan or captured structures (Piper & Deacon 2009).

In addition to the presence of elected councillors, and the opportunity for participation in 'invited spaces' of local governance, residents of Hout Bay enjoy a healthy civil society where there are at least 90 non-governmental organisations that work in the area. These include philanthropic organisations such as Bright Start, religious ones such as James House, issue-based entities such as the National Sea Rescue Institute, and advocacy groups such as the Hout Bay Civic and the environmental group Thrive. Hout Bay also has at least one local newspaper, the *Sentinel*, although this is dominated by 'the view from the suburbs' (Piper, von Lieres, & Anciano 2017), and a number of dedicated Hout Bay sites and groups on social media, such as *Hout Bay Organised* on Facebook. In addition, Hout Bay has a history of popular mobilisation and protest in recent years that is testimony to political resources and energy that exist outside the formal state.

In summary then, residents of Hout Bay have significant democratic access through both the system of formal representative democracy of the ward councillor who sits on the subcouncil and council, and the formal participatory system of the invited spaces of consultation around key policies such as the budget and the IDP. The system of political liberties allows for extensive self-organisation of residents of Hout Bay into civil society formations, mobilisation around key grievances, and a lively public realm through both traditional print media and new forms of social media. In short, the idea that Hout Bay could approach some kind of self-governing republic is not as counter-intuitive as it might first appear. Not only does the City of Cape Town have significant power to shape the future development of Hout Bay, but residents appear to have significant access to this power to govern themselves through the formal systems of democracy.

The main argument of the book is that, in practice, Hout Bay residents do not actually enjoy this power. This is because the system of formal democracy described above is disconnected from key forms of governance that proceed outside the control of the elected representatives and often with undemocratic logics. We make these claims in subsequent chapters. For now, we start with the recognition that today Hout Bay is quite a long way from meeting the conditions for being a republic as it is deeply divided, both socially and politically. It is to these divisions that we now turn.

A republic divided by racialised settlements

As noted in the Introduction, Hout Bay was first an agricultural area before being zoned as 'white' and made increasingly residential from the 1950s. By the mid-1980s residential growth and associated service industries in Hout Bay were on the rise and, as suggested by Figure 1.3, have continued on a steep upward path until the present.

The rapid growth of population in Hout Bay since the mid-1980s is not exceptional in the city, but it is faster than most areas in Cape Town. As Todes (2015: 20) notes, urbanisation in South Africa is driven by the combined impact of economic growth, almost exclusively located in the cities, and declines in rural livelihoods and displacement from commercial agriculture. While this process peaked in South Africa between the 1970s and the 1990s, in Cape Town urbanisation has not slowed. Thus, according to the City of Cape Town (CCT 2014: 14), 'the population grew by 46 per cent in the 15-year period between 1996 and 2011'. By 2016, then Cape Town Mayor Patricia de Lille described the city as 'buckling under the pressure of urbanisation',

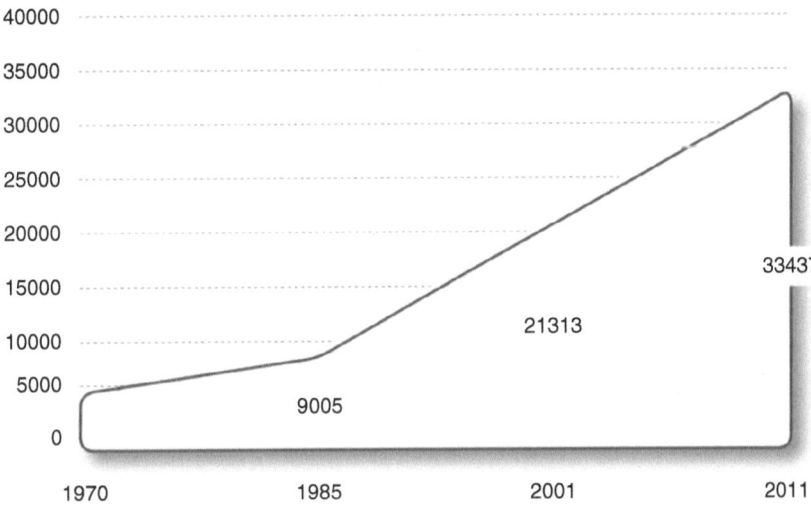

Figure 1.3 Hout Bay population, Census 1970 – Census 2011

adding that while Johannesburg was South Africa's largest city, Cape Town was fast catching up (Sesant 2016).

Notably, the population growth of Hout Bay far exceeds the growth of Cape Town more widely. Thus, Cape Town grew by 3 per cent a year from 1996 to 2011 while Hout Bay grew by 10 per cent a year over the same period. According to the 2001 and 2011 censuses, Hout Bay grew 57 per cent or 5.7 per cent a year, again at a rate higher than Cape Town as a whole. Further, there are significant racial and spatial patterns to the growth of Hout Bay. Colonial and apartheid governments differentiated between European settlers and indigenous groups; the 1949 Population Registration Act required all residents to identify by race as 'white', 'black', or 'coloured' (mixed race). Later 'indian' South Africans were added as a fourth racial category. These four categories were reproduced by post-apartheid rule to provide a basis for affirmation action, land redistribution, and economic empowerment programmes, and so racial identities remain widely used. They are still the primary basis of social identification in everyday life in South Africa, including Hout Bay.

As illustrated by Figure 1.4 there are clearly differential rates of population growth by race group in Hout Bay down time. In 1970, coloured residents were the largest group, but this group has grown the slowest so that today it is smaller than both black and white resident groups. Further, while white residents formed the largest group in the late 1980s, black residents became the largest by 2001 and were clearly so by 2011. The population of Hout Bay by race is quite similar to the rest of Cape Town, with the exception that there are too many white residents versus coloured residents compared to the rest

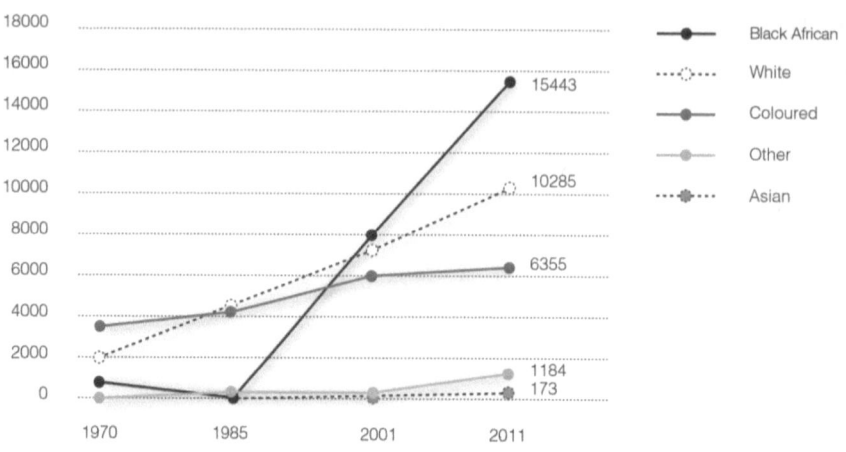

Figure 1.4 Hout Bay absolute population by race 1970–2011

of the city. Nevertheless, in both places the major groups are all represented in roughly the appropriate ratio with no one group forming an overwhelming majority.

In addition to being constituted by three racial groups, the population of Hout Bay is also spatially and socio-economically divided into the three areas indicated in Map 1.2. Almost 100 per cent of white residents live in 'the Valley', which at around 950 hectares constitutes most of the land area in Hout Bay, and over 90 per cent of the residential land. A high majority of coloured people live in Hangberg and the Heights, although a few wealthier families do live in the Valley. Together Hangberg and the Heights are 47 hectares, or less than 5 per cent of Hout Bay's residential land. Lastly, the vast majority of black African residents, including migrants from the rest of Africa, live in Imizamo Yethu on 43 hectares of land, if you include the green belt and informal settlements. This constitutes around 4 per cent of residential land in Hout Bay.

Thus not only are the racial groups who live in Hout Bay segregated spatially, they are also profoundly unequal in socio-economic terms. As reflected in Table 1.1, this amounts to huge differences in settlement type with half the population of Hout Bay, who are black and poor, crammed into mostly informal dwellings in Imizamo Yethu, whereas a third of the population, wealthy and white, is spread over 90 per cent of the upmarket residential area. Thus spatial inequality maps onto a whole range of socio-economic inequalities too, including housing, education, employment, and basic services such as sanitation, water, and the like.[2] Indeed, it is hard to see how Hout Bay could be more divided than it currently is, with three racially, socio-economically, and culturally distinct groups segregated into spatially adjacent places.

Hopefully by now it is clear that Hout Bay is not one community so much as three distinct social groups segregated into three adjacent places. In addition,

Map 1.2 Map of key settlements of Hout Bay

there are areas such as the business district and the harbour (see Map 1.2) that all groups can and do access, but most of Hout Bay is de facto a segregated residential place, colloquially called a 'community' or 'suburb'. To better understand the reasons for this segregation we now explore briefly the history of each settlement under Union and Apartheid rule, and then into the post-apartheid present. We begin with the largest place in Hout Bay, the Valley.

The Valley

In colonial times Hout Bay was settled by Dutch and other immigrant European farmers and was zoned as a white area under Union and Apartheid governments, in particular with the Group Areas Act of 1950. This was despite

Table 1.1 Neighbourhood indicators (Census 2011)

Neighbourhood	2011 Population	Spatial area*	Density*	Sex ratio (M:F)	Higher Ed	Formal housing	Piped water	Electricity
Hangberg & the Heights	+/_ 6000	47 Hectares	12 766 per km^2	92.6:100	45.8%	94%	96.5%	99.2%
The Valley	+/_ 10200	953 Hectares	1 070 per km^2					
Imizamo Yethu	15400	42 Hectares	36 667 per km^2	121.5:100	2.6%	23.2%	25%	80%

* Own calculations from 2011 Census data

the fact that, from the early part of the twentieth century, the area included a majority coloured population working on the farms and in the fishing industry. The only exception to this was the designation of Hangberg as a coloured area in the 1950s, and the forced removal of coloured people living in other parts of Hout Bay to Hangberg. Indeed, it was only with the advent of Imizamo Yethu in the early 1990s that this pattern of settlement changed to any degree.

The major socio-economic changes in Hout Bay began in the 1960s and 1970s when former farmland was sold off for residential development, and the Valley grew into a white suburb. This process of 'residentialisation' saw a significant increase in the population of Hout Bay, reflected in Figure 1.3, a process that continues today. With the advent of democracy in 1994, the opportunity to redress the legal segregation of apartheid arose, and indeed key pieces of racially discriminatory legislation were removed. However, in reality segregation has continued in Hout Bay with little change in settlement patterns. A key reason for this was, on the one hand, the reluctance of the state to pursue an aggressive policy of land expropriation, and on the other, the extension of the market in private residential property, and its opening up to international buyers.

In the case of Hout Bay, this policy approach meant that not only could residents of the Valley retain their land, but also that coloured and black African residents could acquire it only through market mechanisms. Of course, the vast majority lack the financial means to do so, particularly due to the racially framed property owning and job reservation laws of apartheid. Furthermore, the process of applying for land restitution for coloured people forcibly removed in Hout Bay has dragged on for 20 years, with no tangible outcome (PMG 2010). At the same time, though, a number of wealthy coloured and black families have bought property in the Valley, but these are both very small in number and almost exclusively newcomers rather than long-standing residents of Hout Bay.

Hangberg

Before the 1950s, Hangberg was the site of a workers' hostel, and was home to workers categorised as 'African'. Lumber and farming were the main industries in Hout Bay at the time (SAHistoryOnline 2015). In the 1950s, the lumber industry was in decline and fishing, as well as factories processing fish, soon became the largest industry in the area. Coloured people made up the bulk of the labour force and were needed to work in the harbour. At the time, most of the coloured people working in these industries lived in the Hout Bay valley or close to the village.

With the Group Areas Act of 1950, Hout Bay was designated a 'white area', and South Africans who were not white were only allowed into Hout Bay to work as manual or semi-skilled labour. This posed a profound problem for many long-standing coloured residents. Some parts of the apartheid government wanted to move the people categorised as coloured to the Cape flats, but the white owners of the fishing industry wanted a labour reserve close to the harbour. Hence, Hangberg was designated a coloured area to provide labour for the fishing industry (SAHistoryOnline 2015).

The implementation of the Group Areas Act in Hout Bay began a process of moving people classified as coloured from the Hout Bay village to Hangberg. From 1950 to 1980 there was a series of forced removals, and some people were moved twice, first closer to the harbour around Beach road and then into Hangberg. Council houses and flats were built by the state to accommodate people forcibly moved into the settlement. The subdivision of farms due to growing urbanisation of the white residential area of the Valley displaced a large number of labourers, most of whom were coloured, and whose families had long resided there. These became the 'traditional squatters' who lived in informal settlements in Hangberg and on the Disa River (Zille 1990 in Gawith & Sowman 1992).

With a booming fishing industry in the 1970s, the original council houses and flats were quickly overcrowded, and some residents began building their own informal structures. More commonly termed 'shacks' in South Africa, the local residents of Hangberg colloquially termed them 'bungalows'. Initially, most of the informal structures were built in between apartment buildings and behind houses as 'backyard dwellings'. By the turn of the century, however, overcrowding in Hangberg meant that many families were living in informal structures at the top of the settlement, and were encroaching on land that is formally part of the Table Mountain National Park, run by the national government (Ehebrecht 2014). It was the attempt to resettle this community off this land and into formal housing that led to the 'battle of Hangberg', discussed further in Chapter 4.

Imizamo Yethu

In the 1950s, the need for labour in the harbour attracted black African migrant workers precluded by apartheid legislation from ownership or secure leases. Squatting occurred sporadically by pockets of people for more than 50 years (SAHistoryOnline 2015). Initially the effects were minimal on the existing white community as squatting occurred along the Disa River banks and in the backyards of corporate accommodation supplied by the fishing industry. However, by late 1990 more than 2,000 people lived in five main informal settlements, the largest being known as 'Princess Bush' and 'Sea Products' near the Hout Bay harbour. Other smaller settlements were developed at 'Disa River', 'Blue Valley' and 'Dawids Kraal' (SAHistoryOnline 2015). Among these informal settlers were black Angolans and Namibians who had been recruited into the apartheid-era defence force, in particular 32 Battalion, and who were now working in South Africa.

Imizamo Yethu (literally 'our struggle' or 'our collective efforts') was formalised in 1991 in an attempt to consolidate and upgrade the five informal settlements by moving them onto a piece of land belonging to the then Regional Services Council (Carney 2003: 1). Harte, Hastings and Childs (n.d.) state that the identified land was 34 hectares of which 18 hectares were allocated for housing, while 16 hectares were allocated to be developed for community facilities. This suggests that right from the start the idea was to develop place-specific (and therefore race-specific) facilities for Imizamo Yethu. While the initial land was intended to accommodate some 455 families, the number grew quickly to over 3,000 structures by 2003 (Bedderson 2004: 1) with an estimated 7874 residents (Rangasami & Gird 2007: 3). By 2011, the official census estimated that Imizamo Yethu held over 15,000 people in over 1800 formal and 5,000 informal dwellings.

From being the smallest of the three settlements in Hout Bay in 1991, Imizamo Yethu is the largest by some distance today, housing more than half of all the residents of Hout Bay. Thus, the greatest increase of any group of residents in Hout Bay is the 2415 per cent increase in the black population between 1985 and 2001, the period that saw the establishment of Imizamo Yethu. In the last 10 years, an important contributor to population growth has been the influx of significant numbers of migrants from the rest of Africa, especially Malawi, the Democratic Republic of the Congo (DRC), and Zimbabwe. Local estimates put the figure of foreign migrants as high as 40 per cent of Imizamo Yethu residents, but a 2011 survey we conducted of 300 randomly sampled households put the figure at around 15 per cent on multiple measures.[3]

Notably, this survey finding causes us some consternation as it runs against the received wisdom of many who live in Imizamo Yethu and who point to the prominence of Somalis in spaza shops, Namibians in taverns, the Chinese shop, and so on, as evidence for the claim that around 40 per cent of residents are foreign. However, our findings are closer to the 2011 census for ward 74, which identifies 3.3 per cent of the population as 'other'. Assuming the vast majority of these live in Imizamo Yethu this would be about 7 per cent of Imizamo Yethu – roughly half what we found. In addition, a 2003 survey found that 5 per cent of Imizamo Yethu residents were foreign nationals, so this is a threefold increase in 8 years in the settlement (DAG 2003). Although this uncertainty around the numbers of African migrants in Imizamo Yethu is unresolvable without more careful empirical research, the available evidence suggests that it is probably closer to the 20 per cent mark (4,000 people) than the 40 per cent one (8,000 people) often invoked in public forums.

In addition to shedding light on the size and profile of residents in Imizamo Yethu, our 2011 survey revealed that 54 per cent of residents previously lived in the Eastern Cape, with 22 per cent moving from somewhere else in Cape Town and 13 per cent from elsewhere on the continent. Notably, as reflected in Figure 1.5, the patterns of migration have not been even down time, with most Cape Town residents coming early on in the life of Imizamo Yethu, Eastern Cape migrants dominating the 2000s, and more African migrants coming from the mid-2000s onwards. This pattern is important as it speaks to the growing dynamic of xenophobic politics covered in Chapter 6. In addition, however, it casts more light on the social ties that bind in Imizamo Yethu. The 2003 survey found that 40 per cent of Imizamo Yethu residents came from one district in the Eastern Cape. A 2005 survey found that 38 per cent were

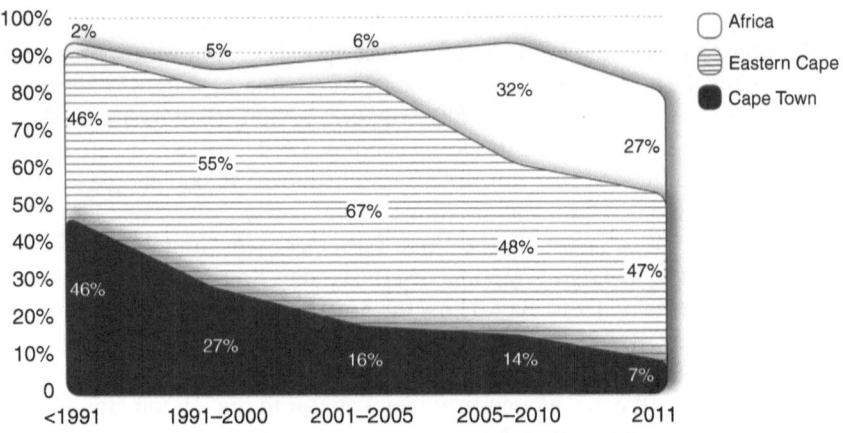

Figure 1.5 Previous residence of Imizamo Yethu migrants 1991–2011

from the Willowvale/Gatyana area (DAG 2003; SALDRU 2006). The 2011 survey found that while 54 per cent of respondents were from the Eastern Cape, they came from all parts, with no one region dominating.

A republic divided by party politics

Hopefully we have demonstrated how, over the last one hundred years, Hout Bay has evolved from one agricultural area into three adjacent but segregated residential neighbourhoods. Considered together, the neo-apartheid coincidence of race, class and even culture that distinguish the Valley, Hangberg, and Imizamo Yethu constitute deep social divides that undermine any easy notion of one local community, especially when given the conflictual history of South Africa. Not surprisingly, then, the idea of a shared political identity is not widely held in Hout Bay, but rather local politics reflects the zero-sum logic of race politics nationally. In this section, we focus on one aspect of this, the rivalry between political parties.

In respect of voting patterns, we have accessed election results down to voting station level for the local government elections of 2000, 2006, 2011, and 2016, and the national elections of 1999, 2004, 2009, and 2014. These we have reconstructed into a set of election outcomes for Hout Bay by including the five voting districts that fall within the place of Hout Bay rather than using the results for Ward 74, which also includes voting districts outside of Hout Bay on the Atlantic seaboard. We also cite the results from the proportional representation (PR) vote at the local level, rather than the ward, as the difference is minimal. Further, we cite the results for the national rather than the provincial ballot for the national elections for the same reason.

Three key trends are evident from the election result. First, regardless of whether local or national elections, the contest is dominated by the rivalry between the African National Congress (ANC), the ruling party nationally, and the Democratic Alliance (DA), the official opposition nationally. Indeed, this contest is a close one, and is intensifying in the sense of pushing out other parties from a 'two horse race'. Second, the party rivalry maps neatly onto segregated place in that the Valley always votes overwhelmingly for the DA, and Imizamo Yethu votes overwhelmingly for the ANC. Over the years, only Hangberg has proved ambivalent, and has moved from mostly supporting the ANC to mostly supporting the DA. Third, both the emergence of a two-party race, the profile of voters, and the swing of coloured supporters to the DA, all reflect larger provincial and national voting trends. Increasingly, it seems that for Hout Bay voters, local elections serve as a referendum on national politics, rather than being driven by local political concerns.

DA/ANC dominance

As is evident from the results of both local government elections and national elections (Figure 1.6), all elections in Hout Bay are about the ANC versus the DA, and this is usually a close contest. In the local elections, the DA's initial success in 2000 was narrowed in the subsequent two elections until 2016. In national elections, the ANC's early success changed after 2009 due to the switch in allegiance of voters in Hangberg. Further, both parties have increased their combined vote significantly in both local and national elections to over 90 per cent, suggesting that electoral politics is increasingly a two-horse race. Notably, this is even more so for local elections than national elections.

Party allegiance maps onto racialised place

There are six voting districts in Hout Bay, three of which are located in what are largely 'white' areas (Llandudno Primary School, Moravian Oranjekloof

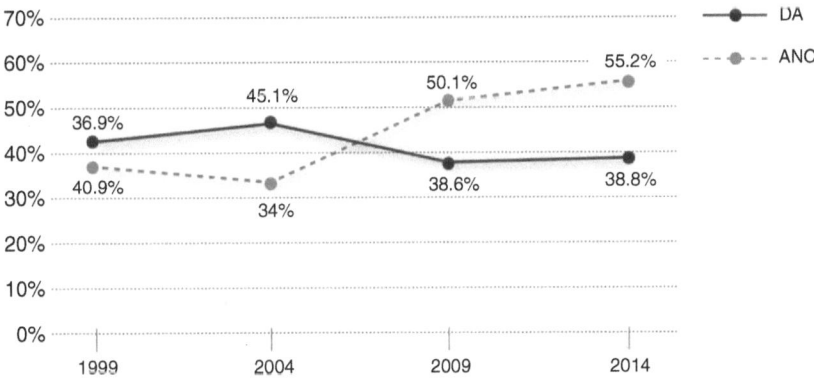

Figure 1.6 National government election results % Hout Bay 1999–2014

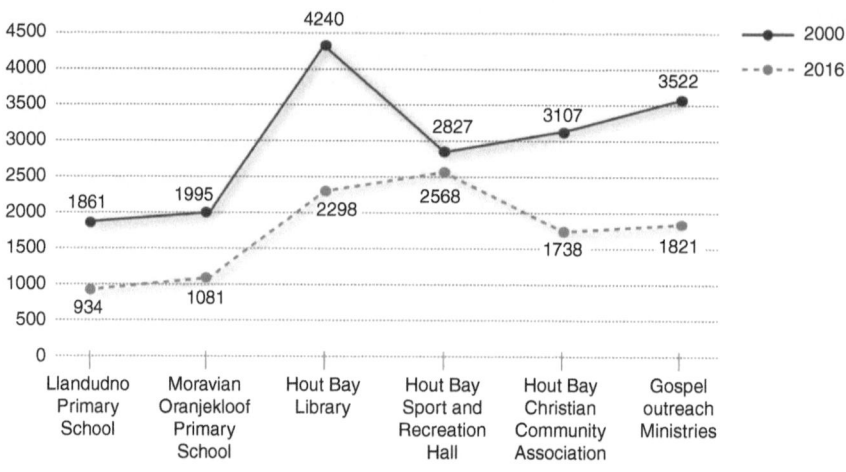

Figure 1.7 Absolute registrations by voting district 2000 versus 2016

Primary, and the Hout Bay Library), one in Hangberg (the Hout Bay Sport and Recreation hall), and two in Imizamo Yethu (Hout Bay Christian Community Association and Gospel Outreach Ministries). Notably, the voting at five of these has been consistent across four local and four national elections, with voting districts in white areas returning overwhelming DA support, and voting districts in Imizamo Yethu returning overwhelming ANC support. The only changeable area has been Hangberg.

Further, if one explores registrations at these voting stations down time, as in Figure 1.7, it is clear that most new registrations are occurring in the white, middle-class area of the Hout Bay Library, and the two Imizamo Yethu voting districts. While this explains the enduring closeness of the battle between the two parties, it also helps to explain their increasing vote share, as the core constituencies of the DA and ANC are both growing. Lastly, as illustrated by Figure 1.8, a key advantage of the DA is the higher turnout of its supporters consistently down time, and across local and national elections. In effect, then, mobilisation at local election time gives it a roughly 5 per cent advantage over the ANC.

Hout Bay local elections reflect national trends

Finally, almost all the major trends in elections in Hout Bay, from the emergence of a two-party race through the profile of voters to the swing of coloured supporters to the DA, are also the key trends in provincial and national elections (Friedman 2009; Africa 2010; Schulz-Herzenberg & Southall 2014). Most political scientists are reluctant to compare figures from local and national government elections as they are typically assumed to reflect different dynamics, in particular the impact of localised issues, actors, and parties on voter choice. However, in South Africa we seem to be witnessing

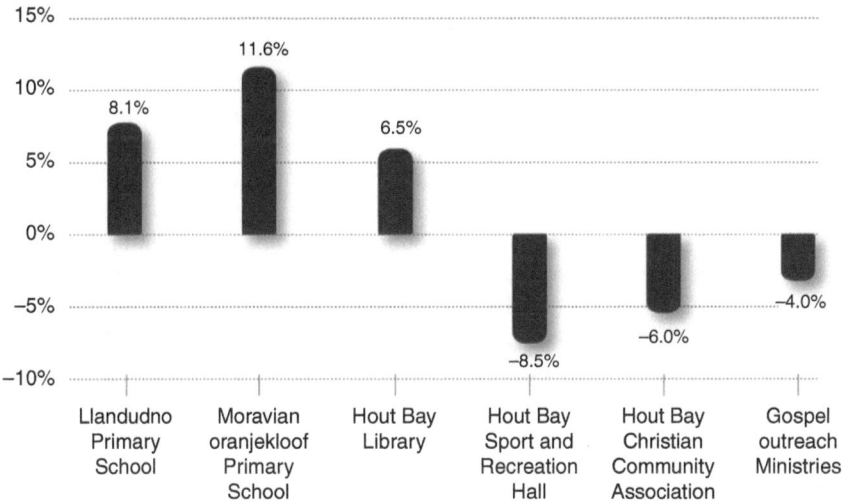

Figure 1.8 Average voter turnout by voter station versus Hout Bay average 2000–2016

the nationalisation of local elections that are increasingly taken as plebiscites on national politics (Piper 2012).

Finally, the fact that local elections in Hout Bay seem more like decentralised national elections raises the question of to what extent they are informed by local issues, events or people at all. Are voters choosing a ward councillor who they think will advance the interests of the place that they live, or are they addressing national debates, issues, and leaders via local proxies? If so, what does this say about the cognitive and practical connection between local elections and local politics? As one of our respondents put it, 'it doesn't matter who the DA put up in Hout Bay, they will get voted in because it's a vote against Zuma and the ANC' (Political representative 2 2017). If accurate, and the electoral patterns suggest that it is, the nationalisation of local elections means that, contrary to their intention, they become largely disconnected from local politics and local political leaders.

Conclusion: a house divided

However dilapidated the 'welcome to the Republic of Hout Bay' sign, it is in better shape than the 'Republic' itself. While there are grounds for believing that the City of Cape Town has the capacity to shape its future development in substantial ways, and that local democracy can enable residents to substantially influence their future, Hout Bay is yet to realise this promise. Today, in both social and political terms, the Republic does not exist. Instead, Hout Bay is divided into three adjacent but segregated places that are deeply divided by race, socio-economic status and politics. Further, as we will unpack in subsequent chapters, the key urbanisms associated with the Valley, Hangberg and Imizamo

Yethu are profoundly different. Belonging and being in Hout Bay means radically different things to different people, depending on where and how they live. Perhaps not surprisingly, these social cleavages are manifested in support for competing political parties, with most white and coloured residents voting for the DA, and almost all black South African residents voting for the ANC. Indeed, local elections have come to reflect national political concerns and judgements and thus, in this way at least, are disconnected from local issues.

Despite the challenges that face Hout Bay, it has many strengths too. It has a vibrant and extensive civil society, it is much better resourced than most areas, and neighbourhoods exist in close proximity, facilitating both an awareness of issues and easy access in addressing them. There are also important social ties that cut across the three communities, principally economic relations of employment, but also personal relations that follow from these too, as well as a shared consciousness of the interdependency of living together in Hout Bay. It is unlikely that many people really think Hout Bay can be made safer, or be socially uplifted, without working across all communities. There are episodic attempts to do this, with varying degrees of success, as will be discussed in Chapter 2. All this means that were Hout Bay residents able to harness these strengths behind a genuinely shared endeavour, there is an excellent chance of success. For this to happen, however, popular agency and democratic desire need to be reconnected with the forces that govern the settlement and shape these decision-making processes in more inclusive and democratic ways.

Notes

1 From www.houtbay.org/2011/09/brand-hout-bay.html.
2 Unfortunately, some of the census 2011 data for the Valley and Hangberg are categorised together, but nevertheless the contrast with Imizamo Yethu remains profound, especially in terms of education levels and services.
3 This survey was conducted as part of an ESRC/DFID-funded project (RES-167-25-0481) entitled 'Agency and Governance in Contexts of Civil Conflict'.

References

Africa, C. (2010). Party Support and Voter Behaviour in the Western Cape: Trends and Patterns Since 1994. *Journal of African Elections*, 9(2), 5–31.
Ajam, T. (2014). Intergovernmental Fiscal Relations in South Africa. In Bhorat, H., Hirsch, A., Kanbur, R., & Ncube, M. (Eds.), *The Oxford Companion to the Economics of South Africa*. Oxford: Oxford University Press, 127–133.
Bagui, L., & Bytheway, A. (2013). Exploring E-Participation in the City of Cape Town. *The Journal of Community Informatics*, 9(4). www.ci-journal.net/index.php/ciej/article/view/982/1052. Accessed 3 September 2017.
Bedderson, S. 2004. The History of Imizamo Yethu. Memorandum to the Executive Mayor. Development and Infrastructure, Public Housing.
Butler, A. (2005). How Democratic is the African National Congress? *Journal of Southern African Studies*, 31(4), 719–736.

Carney, R. (2003). Imizamo Yethu Improvement Project Status Quo as at 30 June 2003. City of Cape Town.

CCT (City of Cape Town). (2014). Available at: https://resource.capetown.gov.za/documentcentre/Documents/City%20research%20reports%20and%20review/Annual%20Report%202012-13.pdf.

CCT. (2017). *City of Cape Town 2017/18 Adjustments Budget 24 August 2017. Annexure A*. www.capetown.gov.za/Family%20and%20home/meet-the-city/the-city-budget/the-citys-budget-2017–2018. Accessed 3 September 2017.

CCT. (n.d.). (a) *Subcouncils*. www.capetown.gov.za/Family%20and%20home/Meet-the-city/City-Council/subcouncils. Accessed 8 October 2017.

Development Action Group. 2003. Imizamo Yethu Survey 2003. www.dag.org.za/images/pdf/researchreports/2003_Report_Imizamo%20Yethu%20Survey.pdf. Accessed 6 August 2015.

DPLG/GTZ. (2005). Having your Say: A Handbook for Ward Committees. Pretoria

Ehebrecht, D. (2014). The challenge of informal settlement upgrading: breaking new ground in Hangberg Cape Town? Master's thesis, Universitätsverlag Potsdam, Potsdam.

Friedman, S. (2009). An Accidental Advance? South Africa's 2009 Elections. *Journal of Democracy*, *20*(4), 108–122.

Gawith, M., & Sowman, M. (1992). Informal Settlements in Hout Bay: A Brief History and Review of Socio-Demographic Trends 1989–1991. *EEU Report 10/92/92*, University of Cape Town, Cape Town.

Harte, W., Hastings, P., & Childs, I. (2006). Community Politics: A Factor Eroding Hazard Resilience in a Disadvantaged Community, Imizamo Yethu, South Africa. Queensland University of Technology. Paper Presented to the Social Change in the 21st Century Conference.

Jolobe, Z. (2010). The Democratic Alliance: Consolidating the Official Opposition. In Southall, R. (Ed.), *Zunami! The 2009 South African Election*. Pretoria: Jacana Media, 131–146.

Local Government. (2000). *Local Government: Municipal Systems Act 32 of 2000*. www.gov.za/documents/local-government-municipal-systems-act. Accessed 25 April 2016.

Njenga, T. M. (2009). From a Thesis: A Critical Analysis of Public Participation in the Integrated Development Plans (IDP) of Selected Municipalities in Some Provinces (Gauteng, Eastern Cape, KwaZulu-Natal and Western Cape) in South Africa. Submitted in partial fulfilment of the requirements for the degree of Master of Social Sciences (Policy and Development Studies) in the Faculty of Humanities, Development and Social Sciences in the University of KwaZulu-Natal, Pietermaritzburg.

Penderis, S., & Tapscott, C. (2014). The Establishment of a Development Local State in South Africa: Between Rhetoric and Reality. University of the Western Cape. A paper presented to the 23rd World Congress of Political Science, Montreal, Canada, 19–24 July.

Piper, L. (2012). Further from the People – Bipartisan Nationalisation Thwarting the Electoral System. In Booysen, S. (Ed.), *Local Elections in South Africa: Parties, People, Politics*. Stellenbosch: SUN Press, 31–44.

Piper, L., & Deacon, R. (2009). Too Dependent to Participate: Ward Committees and Local Democratisation in South Africa. *Local Government Studies*, *35*(4), 415–433.

Piper, L. von Lieres B., & Anciano, F. (2017). The Tale of Two Publics: Media, Political Representation and Citizenship in Hout Bay, Cape Town. In Garman, A., and Wasserman, H. (Eds.), *Media and Citizenship: Between Marginalisation and Participation*. Pretoria: HSRC Press, 120–138.

PMG (Parliamentary Monitoring Group). (2010). Petition on Complaints about Land Claims on Behalf of Western Cape Land Restitution Group: Update. *Meeting of the Rural Development and Land Reform Committee*. https://pmg.org.za/committee-meeting/11777/. Accessed 23 October 2016.

Rangasami, J., & Gird, A. (2007). Rapid Impact Assessment of NMTT's work in Imizamo Yethu, Cape Town, from 2003–2006. Impact Consulting.

RSA Constitution. (1996). The Constitution of the Republic of South Africa, Act 108 of 1996. www.acts.co.za/constitution_of_/index.html. Accessed 3 September 2016.

SAHistoryOnline. (2015). *Hangberg*. www.sahistory.org.za/place/hangberg. Accessed 23 October 2016.

SALDRU (Southern Africa Labour and Development Research Unit). (2006). Migration to Two Neighbourhoods in the Suburb of Hout Bay, Cape Town, 2005. *Survey Report and Baseline Information*. SALDRU, School of Economics and CARE, School of Management Studies, University of Cape Town. www.capetown.gov.za/en/stats/CityReports/.../Migration_in_Hout_Bay.pdf. 23 October 2016.

Schulz-Herzenberg, C., & Southall, R. (Eds.). (2014). *Election 2014: The Campaigns, Results and Future Prospects*. Johannesburg: Jacana and KAS.

Sesant, S. (2016). Cape Town Buckling under the Pressure of Urbanisation. *Eyewitness News*. http://ewn.co.za/2016/05/30/Cape-Town-buckling-under-pressure-of-urbanisation. Accessed 23 October 2016.

SPLUMA. (2013). *Spatial Planning and Land Use Management Act 16 of 2013*. www.sacplan.org.za/documents/SpatialPlanningandLandUseManagementAct2013Act16of2013.pdf. Accessed 18 September 2017.

Todes, A. (2015). The External and Internal Context for Post-Apartheid Urban Governance. In Haferburg, C., & Huchzermeyer, M. (Eds.), *Urban Governance in Post-Apartheid Cities: Modes of Engagement in South Africa's Metropoles*. University of KwaZulu-Natal Press.

White, F. M. (2008). Strengthening democracy: the role of social movements as agents of civil society in post-apartheid South Africa. Doctoral dissertation, University of London.

Winter, K. (2017). What's Driving Cape Town's Water Insecurity, and What Can Be Done About It. *The Conversation*. https://theconversation.com/whats-driving-cape-towns-water-insecurity-and-what-can-be-done-about-it-81845. Accessed 16 May 2018.

World Bank. (2003). *World Development Report 2004: Making Services Work for Poor People*. World Bank. https://openknowledge.worldbank.org/handle/10986/5986. Accessed 18 September 2017.

Interviews

1. City official 1. (2015). Interviewed by Laurence Piper. 7 April 2015.
2. Political Representative 2. (2017). Interviewed by Fiona Anciano. 18 October 2017.
3. PPU. (2016). Public Participation Unit, City of Cape Town. Interviewed by Laurence Piper. 23 March 2016.

2 A river of grime

Governing water and waste

On a Thursday night in the Hout Bay Museum, an eclectic group of locals have gathered for a workshop on how to tackle water, waste, and land problems in Hout Bay. 'What is the biggest challenge facing Hout Bay's environment?' shouts the facilitator to the mixed crowd. 'Poor sanitation in the township' replies a women from Imizamo Yethu, and several residents shuffle towards her. 'The dirty river' responds a gentleman from the Valley, soon surrounded by numerous others. And so begins the 'soft shoe shuffle' exercise led by the development organisation *in/formal south*.

The aim of the exercise is for participants to move to the person who shouts out the challenge that they identify with the most. After ten minutes of shouts and shuffles it becomes clear that almost every one of the diverse participants agrees on a similar and linked cause of environmental problems in Hout Bay, and so all residents end up in one large group. This process of participation highlights how, regardless if you are from the wealthy valley or a poor township, you can agree on the challenges that face Hout Bay. It is also reasonable to assume that, if such a diverse group could agree on a common environmental challenge through a participatory process, this process should have led to a policy change improving the environment of Hout Bay. And yet, as this chapter will show, it did not.

This chapter addresses concerns about water and waste in Hout Bay, specifically focusing on the Disa River that flows through the 'Republic'. It presents a case study of a civil society-led participatory process that tried to resolve Hout Bay-wide environmental problems through rehabilitating the river. Remarkably, despite generating agreement between the diverse residents of Hout Bay, and in-principle support from City line departments and local politicians, the project failed to get the state support necessary for implementation. As we show in what follows, the problem was the disjuncture between the holistic design of the project and the compartmentalised logic of City bureaucracy.

In effect, the *in/formalsouth* environmental project was a round democratic peg and the City bureaucracy a square bureaucratic hole. Without political support from the highest levels of City leadership to facilitate co-operation across departmental silos, cultures, spheres, and scales, the bureaucracy could not 'see' or 'do' the *in/formalsouth* project without running up against its own rules and hierarchies, and the institutions of local democracy did not have the authority to make the bureaucracy work differently. This conclusion is reinforced by the consideration of a second state–society project on the Disa River, which was more successful precisely because it was able to translate its objectives into terms visible to the line departments of City Hall. In short, this chapter reveals the constraints of local bureaucracy on local democracy. When connected to the bureaucratic governance of line departments, local democracy has purchase over City Hall, but when traversing departments and their logics, local democracy is disconnected from bureaucratic power.

Waste and water: cleaning up the 'Republic'

Hout Bay has a relatively unique urban geography. It is bounded by nature, sitting in a valley enclosed on three sides by mountains and on one side by the Atlantic Ocean. Having no directly contiguous urban neighbours creates a rare setting with natural 'buffer' areas on all sides. A positive consequence of this is that Hout Bay should have a higher biodiversity and ecological integrity than many of the other urban areas of Cape Town (*in/formal south* 2015). Yet, Hout Bay faces environmental challenges befitting most urban settings.

In Imizamo Yethu, significant environmental health problems are linked to poor sanitation and solid waste management, leading to grey water pollution, dirty drains, blocked or leaking sewage, insufficient toilet facilities, and piles of rubbish. The mostly informal area in the highest part of Imizamo Yethu, Dontse Yakhe, is infamous for its poor sanitation. As one environmental report noted, Dontse Yahke is 'riddled with open cess-pits of foul effluent' (HBRRA n.d.). In the rest of the township toilet facilities range from a few shared waterborne stations to buckets that are inadequate and often overflow. An Imizamo Yethu community leader describes how 'there is no sanitation, not enough toilets and areas are not serviced' (Community leader 1 2015). Residents are also fearful of walking to toilets at night due to the high crime levels and thus human excreta is left outside shacks in the street. (Residents 2012; Froestad 2005).

A City official explained that buckets used by residents for night soil are emptied into the storm water drains in the morning. To his knowledge there are no 'decanting facilities', that is, containers connected to the sewer where you can empty a full bucket. So, 'at the moment they're decanting into the storm water system, which then comes out at the river'. This results in what is called 'black water' (City official 3 2017). The City did try to deal with sanitation by putting in more toilets in parts of Imizamo Yethu, however these were effectively 'privatised' with certain households putting locks on the toilets to prevent them being broken from overuse (River activist 1 2015). Imizamo Yethu's fragile environment is also compounded by solid waste that is not disposed of properly. The layout of Imizamo Yethu on a steep slope with narrow, congested streets means that City garbage trucks cannot access houses directly. This garbage piles up on street corners and around shipping containers where mixed waste is stored for removal by trucks to landfill. It both contaminates the township and ultimately ends up in the river system (City official 1 2015).

The poor management of waste in Imizamo Yethu has serious health consequences. According to the municipal Environmental Health Department at least 51 per cent of Imizamo Yethu residents fall within a high-risk health category. In this area, the potential is extremely high for diseases to spread rapidly and in an uncontrollable manner (Froestad 2005). The leader of a community health organisation describes how 'children play in rubbish' and that 'with waste you often have rats and we've had children that have been bitten by rats in their sleep'. She explains that it is not possible to definitively link Imizamo Yethu's high levels of diarrhoea to waste and sewerage problems but that there is most likely a correlation between the two (Health Forum leader 2015).

Hangberg too suffers from environmental degradation in the form of solid waste and grey water concerns. Residents are concerned with street level waste and overflowing garbage containers. One resident who runs a cleaning business and has tried to set up recycling explains that Hangberg, 'is just screaming for help and waste is the biggest challenge ... kids ... playing where the bins are ... everyone is in danger the way its filthy now up there. The city has no control' (Waste activist 2015). A City community health worker explained that 'Hangberg is filthy basically, there is solid waste and there are dumping sites everywhere'. While sanitation and water concerns are not as bad in Hangberg as they are in Imizamo Yethu, there are nonetheless problems with illegal sewerage connections which 'puts a lot of strain on the system ... there are always burst pipes'. The official is also concerned that sanitation and waste problems lead to 'huge challenges in term of public health ... it's like you're waiting for cholera or something to break out' (City official 8 2015).

While the Valley faces its own share of environmental concerns, these are centred predominantly around the Disa River, rather than street waste or sanitation problems. The river running through Hout Bay from the lush upper valley, alongside Imizamo Yethu and down to the beach, has been a significant source of conflict over the years. While it offers the potential for a beautiful natural site and public area for all Hout Bay residents, in reality it is 'basically an open sewer canal' (River activist 2 2017) and is claimed to have one of the highest oceanic E. coli[1] counts in South Africa. At one test it contained nine hundred million E. coli bacteria per 100ml water where normal levels are two hundred per 100ml (Bardouleau 2010). This level of E. coli is higher than one would find in normal sewerage (River activist 1 2015). An environmental report noted that, 'highly-elevated bacterial counts ... in the Hout Bay River will contribute substantially to the ecologically unbalanced river ecosystem ... The river is typical of a disturbed system having poor to non-existent biological connectivity' (HBRRA n.d.; see Figure 2.1).

Figure 2.1 Contaminated water flowing into Disa river

The river is so highly contaminated that there is little to no animal life. There was one river otter, but it died in 2017 after getting septicaemia from the polluted water. Recently there were sewerage worms in the river, 'that can survive in very low oxygen environments because they beat their tails to make their own oxygen ... but even they died' (River activist 2 2017). Certainly, the City is aware that the Disa River has a 'water quality' problem and that 'health is the main issue'. It is a challenge that has been 'going on for a long time' (City official 3 2017). The City of Cape Town has at various times over the last decade warned beachgoers not to swim in the sea or the lagoon estuary due to the unsafe levels of E. coli (Cronje 2014).

While most interviewees and researchers attribute the contaminated river and high levels of E. coli to sanitation and waste problems originating in Imizamo Yethu, the historical management of the river is also to blame. During the course of the previous century, the water from the valley catchment area was redirected to feed the City of Cape Town to the north and east of Hout Bay. The river 'lost a significant volume and that fundamentally changed its nature' (River activist 1 2015). A second concern is that, in many parts, the river is bounded on both sides by private landowners. The centre of the river is commonage, but landowners are able to extract water as they wish (although there are legal limits in theory) and this reduces the flow of the river (River activist 1 2015). Nonetheless, any attempt to regenerate the Disa River would need to focus on sanitation and waste concerns in Imizamo Yethu.

Participatory democracy and the cleaning of the river

It is into this environment, described by some as a 'river of shit', that the urban development organisation *in/formal south* has entered. *In/formal south* had been working on issues of river management and biodiversity in other parts of the Western Cape and an environmental activist in Hout Bay had heard of this work and used personal connections to meet with *in/formal south*. This meeting resulted in strategising a potential project to rehabilitate the river and work on related environmental concerns in Hout Bay. Prior to this meeting, *in/formal south* had already set up a partnership with the Dutch funder Cordaid. Cordaid would match funding with *in/formal south* to support projects that the organisation brought to them. *In/formal south* presented the proposition of building a 'Genius of Place' project which would take 'a systems perspective on complex urban challenges' and focus on 'creating green economy opportunities related to water and waste' (Actuality 2017). In early 2015, Cordaid offered funding of around R500,000 (US$34,000) to initiate and set up the project (*in/formal south* staff 1 2017; *in/formal south* staff 2 2017).

A key criteria from Cordaid was that they would only support a project which had written support from the government. The first task of *in/formal south*, then, was to get buy-in from a government actor. Using personal networks, they approached the Cape Town Partnership, a non-profit organisation that has government support to bring actors across the City together to roll out collaborative projects in multiple spheres (CTP 2017). The Cape Town Partnership put *in/formal south* in touch with the nascent Hout Bay Partnership (HBP). The HBP, officially formed in 2014, is modelled on, and incubated by, the Cape Town Partnership. Its goal is to

> be a uniting force behind Hout Bay ... to be a connector and open doors for projects and get people to work together and connect businesses with non-profits, with government, open doors for funding, yeah, just be a connecting, uniting force.
>
> (HBP staff 2015)

It was funded through the DA-run provincial government, with a representative from the Western Cape Province and a representative from the City of Cape Town on the board (HBP staff 2015). The representative from the Province wrote a letter stating that they would support the *in/formal south* project. *In/formal south* took the letter seriously and were optimistic they would have ongoing government support, and potentially even future government funding (*in/formal south* staff 1 2017).

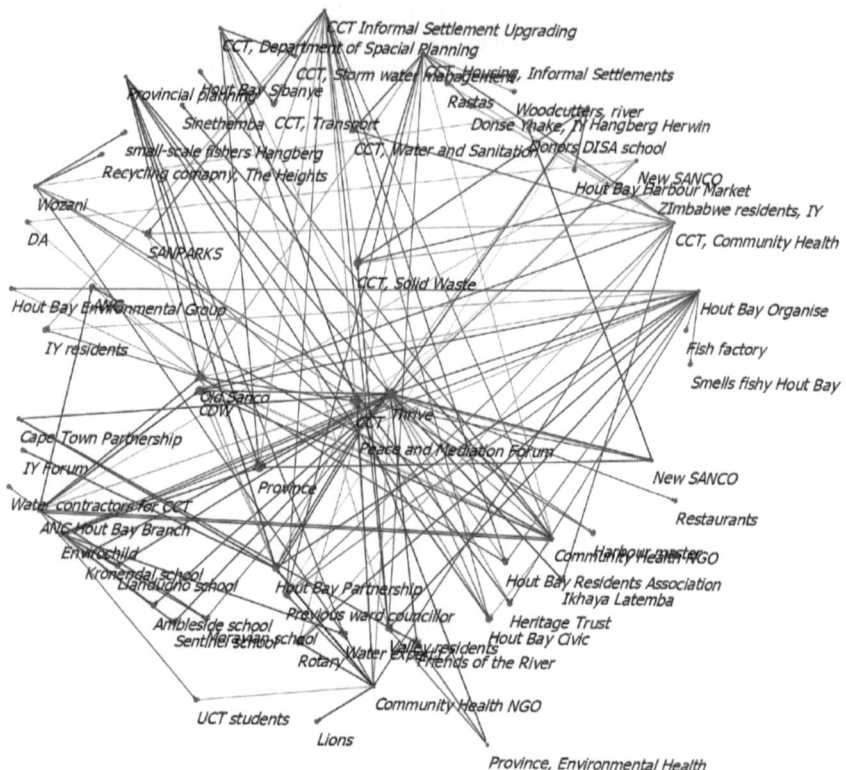

Figure 2.2 Stakeholder mapping Hout Bay (water and waste)

A second of Cordaid's funding criteria was that any project be community-led and build a 'community of change'. A community of change is understood as a group of capacitated agents committed to working on a sustainable development project around the themes of water or waste in a collective and participatory way. While water and waste were identified as key thematic areas, the precise focus, framing, and timing of the project was to be determined collectively by the community of change once constituted. To this end a further organisation, the Sustainable Livelihoods Foundation (SLF), was contracted to conduct a 'stakeholder mapping' exercise (see Figure 2.2). They interviewed over thirty respondents relevant to the management of waste and water in Hout Bay, including civil society and community leaders, government officials and political representatives, and business owners (Piper, Anciano, Sikota, Williams, & Muteti 2015). This allowed *in/formal south* to build a clear analysis of who to include in potential planning regarding environmental challenges in Hout Bay. Alongside, and informed by, this process *in/formal south* ran its own workshops with community leaders and conducted several participatory exercises, such as the 'soft shoe shuffle' described at the start of this chapter.

By mid-2015 *in/formal south* had a clear idea of what the main issues were in Hout Bay related to water and waste management, and who would be most interested in supporting a project to deal with these. The groundwork had been laid, the community mobilised and there was relatively strong buy-in from almost all sectors of Hout Bay. Indeed, the Sustainable Livelihoods Foundation report showed that there were several NGOs as well as the HBP with 'high motivation' to drive a water and waste project. There was less overt support from Imizamo Yethu and Hangberg linked community organisations such as SANCO and the Peace and Mediation Forum (PMF) (see later chapters for details). Nonetheless, Imizamo Yethu residents and civil society leaders had shown an interest in the project and attended several workshops. They indicated they were keen to participate and generate community support if tangible benefits to cleanliness, health, and potential job opportunities would result (Piper et al. 2015). A steering committee to drive the project was established with leaders from all three communities in Hout Bay (Environmental activist 2 2017).

Based on this support and on experience from a successful biomimicry water purifying project in the Langrug informal settlement in rural Western Cape, *informal south* decided to drive the concept of 'living corridors' and biomimicry. Biomimicry is 'an approach to innovation that seeks sustainable solutions to human challenges by emulating nature's time-tested patterns and strategies' (Biomimicry Institute 2017). The idea is that you can create processes to reduce waste, and improve sanitation and river health, by using (green) plant-based cleaning processes rather than traditional (chemical) waste-water treatment works. Hence, *in/formal south* proposed establishing a 'living corridor' of natural, indigenous vegetation that would start in Imizamo Yethu and run down to the river, and in the process it would filter and clean 'black water' and 'grey water'. Allied to this was a proposal for a solid waste recycling/upcycling project that would be staffed by Imizamo Yethu residents and would contribute to job creation. These initiatives were widely supported by community members (*in/formal south* project meeting 2015).

While building a 'community of change' and designing a waste and water treatment project, *in/formal south* also started meeting with City officials and political representatives. They recognised that they would need government support from a technical, legal, and financial perspective, and indeed to meet Cordaid requirements. *In/formal south* decided to focus on government officials rather than political actors, because of experience with previous Cordaid projects that had received Mayoral support and started well, but then collapsed when the Mayor decided she no longer wanted to support the project. So *in/formal south* thought 'if we build it from the bottom up ... create a whole kind of movement around the project ... it won't collapse without political support' (*in/formal south* staff 1 2017).

However, they found the processes of engaging City officials very challenging. It was difficult to know which department to target and work with and to find someone from inside government to champion the process. Through a persistent phone call campaign they were able to set up and present to a handful of officials at several meetings in the City of Cape Town (*in/formal south* staff 2 2017). In these meetings there was generally good verbal support for an integrated water catchment management project based on biomimicry principles, particularly from departments such as Informal Settlements and Environmental Health. One official noted that 'we are supportive ... we try to promote alternative service options' (City official 4 2015). Another explained, 'This project is so welcome. It is like a blessing. Nobody else is going with an innovative idea. They stick to what they know' (City official 5 2015). There was less enthusiasm from utilities departments dealing with waste treatment, but even they stated that 'there is potential' (City official 6 2015). There was an overall sense that the project was a good idea and that planning on the ground and community engagement should continue (City meeting 2015).

In the early phase of the project, *in/formal south* did not focus on political representatives' support, but they did not ignore this avenue of influence either. They realised that enthusiasm from officials would not be enough to take the project forward. They first tried to contact the (then) DA ward councillor, who did not respond personally to any communication. They also approached the ANC candidate standing for ward councillor in the next elections (essentially hedging their bets before the 2016 local government elections). They had more luck with this avenue, and through the ANC were able to present at a ward committee meeting, which was chaired by the (then) DA ward councillor. Many community members attended and, 'we got loads and loads of support ... everybody was on board ... even [the ward councillor] although she wouldn't meet with us separately' (*in/formal south* staff 1 2017).

The ward councillor recommended they escalate the idea to the next level of political decision-making, the subcouncil. Next, *in/formal south* met with the subcouncil chair and presented to him individually. Despite this, the proposed plans for water catchment management were not added to a subcouncil agenda, and thus no political decision to support the project was forthcoming. Notably, a former official working on land in Hout Bay explained how subcouncils are 'biased towards the more upper-income' and that they tend to represent a narrow band of residents because of who sits on the council, meaning that those with personal connections and resources are more able to influence them (Denoon-Stevens 2017). Confronted with this dead end, *in/formal south* then approached the Mayoral Committee Member for Tourism, Events and Economic Development and presented to him; the project again was not escalated further politically, nor put onto any decision-making agenda.

At this point *in/formal south* were starting to get concerned. The officials and their departments could not move forward without a clear political mandate and there was no forthcoming political support. Given that the project had Dutch funding, *in/formal south*, as a last attempt, got the Dutch consulate to arrange a roundtable discussion with senior City officials. This too did not lead to any progress (*in/formal south* staff 1 2017; *in/formal south* staff 2 2017). By late 2015 it was clear the project was not going to succeed. The organisation informed those who had been involved that, 'after a long and tricky road, unfortunately there does not seem to be a lot of immediate buy-in from the city. This is one of the requirements from Cordaid for them to start up and fund projects.' Cordaid funding came to an end, and so did the idea of one cohesive environmentally centred water and waste project for Hout Bay.

Mapping failure

Why did the project fail and what does this tell us about the promise of 'participatory democracy'? There is demonstrably an urgent need for water and waste treatment in all three areas of Hout Bay. There was also substantial buy-in from a large number of civil society leaders in all three areas, notably Imizamo Yethu and the Valley. This was demonstrated not just by the support of the ward committee but also by the consistently high turnout at project meetings, the establishment of a steering committee, and in feedback given in the stakeholder mapping process (Piper et al. 2015). Yet none of this community support from below, coupled with palpable health concerns, was able to influence formal state decision-making.

Participatory democracy through 'invited spaces'

As we note in the Introduction, in many urban settings in the Global South, conventional forms of representative liberal democracy are inadequate for representing residents effectively. The processes of formal democracy, such as the election of a ward councillor, have thus been 'deepened' to include that of participatory democracy, where citizens have direct roles in public choices and engage deeply and regularly with substantive political issues. Officials are, in return, responsive to their views (Cohen & Fung 2004). The practices of participatory democracy relate most clearly to residents of Hout Bay through several legislated forms of public participation outlined in Chapter 1, and reflected in Figure 2.3, including ward committees, subcouncils and the budget and IDP processes. In the *in/formal south* process it was ward committees that held the most potential as a tool of participatory governance. The IDP is only conducted once every 5 years and did not fit with the timelines of the Hout Bay river project.

In respect of the ward committee in Ward 74, the ward councillor described how he 'met individually with all the key NGOs before the selection of the committee' (Political representative 1 2017). Councillors decide on which sectors they want represented on the ward committee and how many people they want to allocate to each sector. In this case, the councillor chose three

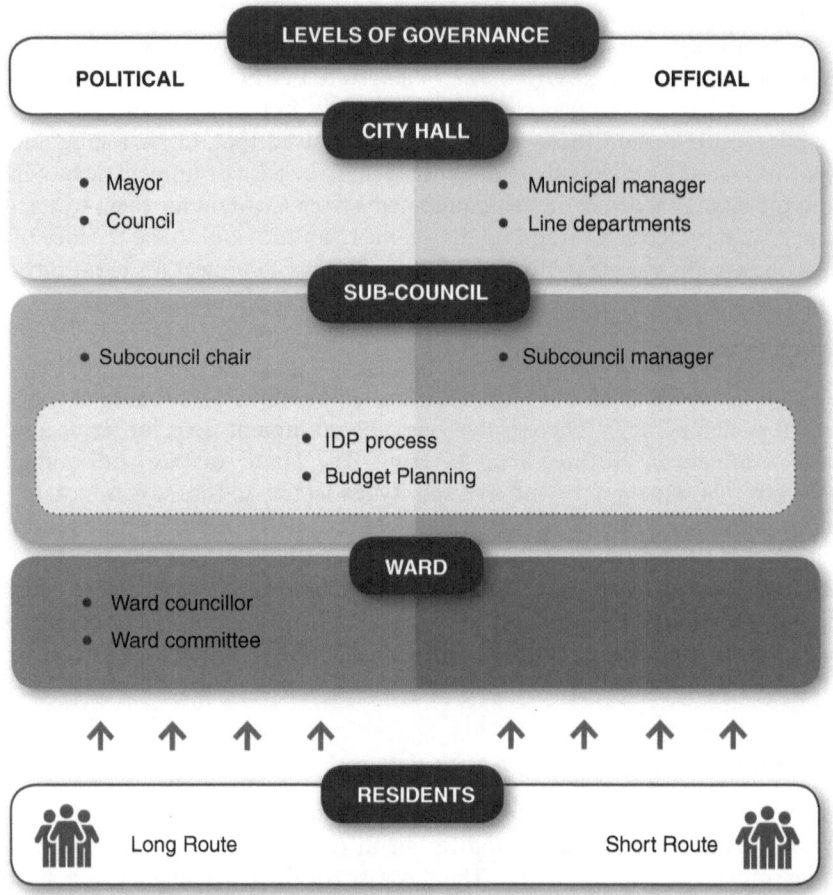

Figure 2.3 City of Cape Town simplified diagram of political and official channels of engagement

seats for safety and security and one for environment. This was because there were three strong organisations in the safety and security sector in his ward and he wanted each to have representation. He felt it was an important concern for the whole ward and he also wanted to minimise contestation. The final selection of the ward committee was not through democratic ballot but by lot. All the names of the organisations contesting each seat were placed in a bag and an official overseeing the process 'literally draws lots'. The councillor had suggested that where there was more than one interested party, 'that they talk it out and decide between themselves'. Nevertheless, for one or two seats there was still competition, decided in the end by whichever name was pulled out of a bag (Political representative 1 2017).

Ward committees have generally been viewed as an ineffective means to deepen democracy. Numerous pieces of research across the country attest to this, illustrating that ward committees do not engage large, representative numbers of citizens, that there are flaws with the design of the committees, and that they can be exclusive partisan spaces (see Barichievy, Piper, & Parker 2005; Piper & Deacon 2008). Research in Cape Town has shown that committees are an avenue for vying for political visibility and leadership positions. Ward committee and subcouncil meetings are open to the public but formal representation and written permission to speak give little meaningful voice to residents (Thompson 2014). Even if ward committees were democratically elected, they still would not link residents' interests with bureaucratic planning systematically. It is not surprising then, that, even though *in/formal south* were able to get significant ward committee support, this did not escalate to meaningful political action.

In design, South Africa's participatory governance system does situate the ward councillor as the first port of call for representing urban residents. A ward councillor should be able to escalate key issues to a City council. In Cape Town this happens through the intermediary level of subcouncils. In the *in/formal south* project, the team were unable to mobilise the ward councillor sufficiently to drive the issue to the subcouncil, and neither could they mobilise the subcouncil. Without a political champion, 'invited' forms of participatory democracy face a brick wall.

Why did the project have no political champion? There are both systemic and individual reasons for this. On the individual level, the ward councillor at the time was disinterested in this project, but arguably in Hout Bay-related projects in general. She was serving out her final year before new elections, for which she was not standing. She had a full-time professional job which constrained her time. She had, as following chapters will show, a complex and challenging relationship with many different groups in Hout Bay on issues ranging from housing to security, and had over the years struggled to reconcile divergent community interests or make meaningful progress on different development projects. This may have contributed to her personal lack of interest. As an official working in housing noted:

> The councillor of the Hout Bay ward does not have a good relationship with informal settlement communities. She never goes to their meetings if she is invited. There is no relationship as far as we are concerned. She may have certain people who report back to her but we don't know who they are. There is no actual real relationship in Hangberg and Imizamo Yethu. I don't know why, perhaps she is afraid of people. People don't know where to find her. We have invited her on numerous occasions but she doesn't come, unless she is tasked on a political level. She will be there when the Mayor is there, under protection. The ratepayers vote for her and maybe they get the quality out of her, but not the informal settlement people.
>
> (City official 5 2015)

In this case, even in the Valley, aligned ratepayers did not 'get the quality out of her'. Beyond personal reasons, there are also systemic reasons why the ward councillor may have been disinterested in a broad-ranging water and waste project. Like most urban centres in South Africa, Hout Bay is faced with conflictual intra- and cross-community dynamics. The enduring racial and class divides of apartheid coupled with the logics of clientelist political resource allocation, found broadly in post-apartheid South Africa (Dawson 2014; Beresford 2015; Anciano 2018), means that the dynamics of representation between the three areas of Hout Bay, and the dynamics of representation within each area, are often conflictual.

Projects involving housing, schools, transport, security, and the building of a clinic, to name a few, have been fraught with inter-community conflict, often leading to protest and challenges for political representatives such as ward councillors. Perhaps for the ward councillor it did not make sense to champion a project that may, like so many others, result in conflict, particularly one she would not see to completion. Indeed, one official working in spatial planning explained that their department had been told by a senior City politician to 'leave Hout Bay alone'. The area was viewed as too troublesome and that any attempt to do a detailed spatial plan would invariably lead to conflict. To date no ward level, detailed spatial plan exists for Hout Bay (City official 4 2015).

Without a ward councillor as a political champion, *in/formal south* were compelled to move up the political chain to the subcouncil, a level of representative government between a City council and a ward. A subcouncil has authority over laws delegated to them by the larger City-wide council. These are fairly wide-ranging and include 'the right to call to account every single line department in the city ... [including] oversight of staff', alongside 'opportunity to comment and give input into IDP processes as well as budget planning and implementation' (Political representative 2 2017). The subcouncil would be able to play an influential role in any project relating to water and waste as they are mandated to 'review and evaluate the needs of the municipality in order of priority' (CCT 2003).

Thus, for *in/formal south*, getting the political support of a subcouncil chairperson would add significant political influence to any developmental project. As with the ward councillor, structural reasons influenced the subcouncil chair, who thus had no incentive to drive the project, or indeed even put it on a subcouncil agenda to be discussed more broadly. Beyond concerns with Hout Bay, any developmental governance project is handled cautiously by the City. As a political representative explained, 'the DA is so terrified of failing a financial audit at any point that there is no fast-tracking of any project. Projects that have not historically been planned for or are large, well every "t" must be crossed ... we cannot afford questioning' (Political representative 1 2017). Analysing the failure of the *in/formal south* project demonstrates that the formal democratic process did not work effectively as a means to connect a civil society-led developmental project to bureaucratic government.

A further concern raised by those who had been involved in the project from the start was that the leaders of *in/formal south* were not Hout Bay residents:

> It was an outside organisation running the project. It had support from all sectors of Hout Bay which was impressive, but the actual driver was not someone based in Hout Bay. Thus, when the funding ran out for salaries, the project died. It was very frustrating.
>
> (River activist 2 2017)

Another internal leader of the project, noted that the fact *in/formal south* were not Hout Bay residents may have been troubling for the City, and would not have assisted in putting pressure on political representatives or City officials (Environmental activist 2 2017). Direct action as a pressure point to complement, or indeed challenge, ineffective invited forms of participatory governance was thus not forthcoming in the *in/formal south* project. Both the invited and invented forms of participatory democracy had failed to yield any movement on a developmental water and waste project for Hout Bay. This demonstrates that political mechanisms of democratic representation are often disconnected from residents' wishes, even where a project has potentially wide support. Indeed the 'disconnected' democratic system did not even allow the issue to reach any decision-making agenda.

Bureaucratic governance

With the 'long route' of political representation failing the *in/formal south* process, the alternative was the 'short'-er route of bureaucratic engagement (World Bank 2004). However, this too failed, for a variety of reasons that we shall discuss. The notion of bureaucratic government essentially involves direct rule by City Hall, where 'government' is used as a verb, as in 'to govern'. Bureaucratic governance refers to the routine activities of the line departments of the City of Cape Town, which work in classic Weberian terms of hierarchies with formal lines of authority, with fixed areas of activity and rigid division of labour, staffed by qualified professionals who make decisions informed by regulations. Thus, every day in Hout Bay the City of Cape Town engages with residents on a range of service delivery issues, from waste removal to tarring the roads. Indeed, this is the core mandate of (local) government. The following sections will unpack five characteristics of bureaucratic rule in Cape Town, namely: (a) departmental silos; (b) mentalities; (c) the restructuring of local government; (d) the relationship between local government departments and other spheres of government; and (e) the impact of scale.

(a) Silos

The City of Cape Town is run along departmental lines, with each department responsible for its own service function. There are 44 departments, key of which include Transport and Urban Development, Urban Integration, and City Health (CCT n.d.). In theory, departments can initiate and run projects that they feel will support their mandate so long as the project falls within legislated boundaries and budgets. As the *in/formal south* process shows, however, anything too large or multi-sectoral is unlikely to get support from a department without the more senior political backing of City leadership. The main reason for this is that departments work as silos. The *in/formal south* team approached a range of departments they felt would be relevant to the project, from Informal Settlements, to Environmental Health, Storm Water Management, Biodiversity, and Waste Management. As *in/formal south* explained

> interdepartmental work was one of the huge challenges, where does this project sit, because it's addressing water, it's addressing informal settlements, it's addressing rivers, it's addressing waste-water treatment, it's addressing economic development, it's addressing social development, so it's like where to go? Our approach was, let's just keep speaking to different groups until it maybe finds a home.
>
> *(in/formal south* staff 1 2017)

It never did find a home. No department felt it fell primarily under their mandate.

In interviews with departmental officials, they would often say they supported the idea but that the project leaders should 'talk to a different department' to see if they could drive the project (Piper et al. 2015; City meeting 2015). An official from the Environmental Health department explained:

> We're at a point now where in the City you're very defined in what you can or can't do. So thinking out of the box doesn't happen. … Getting mobilisation and action from departments is a huge problem, especially when it's outside their core function. … The minute you ask for something more or to be done a different way or, it just kind of doesn't work out.
>
> (City official 8 2015)

This challenge of bureaucratic government is identified by other researchers too, with Millstein quoting a City official: 'what I basically try to say is that the one department does not know what the other department is doing … and sometimes we are confusing the community … we are not working together as a local authority and as one community' (Local official cited in Millstein 2010: 5). As the Hout Bay community health forum reported:

> Our interactions with government structures seemed hopeful at times, but in the end the South African disease of lack of co-ordination among government departments stalled any progress. Let me illustrate this with only one of countless examples: we responded with excitement when

an inter-sectoral team of four government departments was about to be established – Housing, Roads, Cleansing, Health. One meeting took place and this was the end of all further co-ordination.

(Froestad 2005: 351)

Even the previous Mayor of Cape Town and current Western Cape Premier Helen Zille notes that 'bureaucracies are terrible things', difficult to co-ordinate and often wrapped in red tape. Any developmental project, particularly involving land, can take years to deliver (Zille 2017).

(b) Departmental mentalities

There are, of course opportunities for partnerships across departments, and certain departments are designed to manage cross-sectoral programmes, such as Informal Settlements or Spatial Planning. However, as *in/formal south* noted:

Different departments have different kinds of ethos; some are open to new ideas but some wouldn't be used to people like me coming, especially not in an engineering space, maybe the informal settlements guys … *now* I understand that if the City was going do a project like this it would set up a working group and it would put all the departments together and then put out a project call, that may work, but on their own they couldn't respond to our ideas.

(*in/formal south* staff 1 2017)

This finding is echoed in research on environmental policy change in the City of Cape Town. The researchers found that different departments have different 'mentalities'; the Planning and Building Department viewed officials at the Environmental Department as 'activists', while the Environmental Department viewed officials in Planning and Building as 'a bunch of bureaucrats' who erected barriers to protect their processes. This cultural bypass led to the departments finding themselves at loggerheads and the stalling or failure of projects (Froestad, Grimwood, Herbstein, & Shearing 2015). Even where departments are able to work together under the leadership of one department, this type of planning takes years to implement. A city official explained that, 'working on projects that require departmental coordination is very time consuming'. The spatial plan for the district in which Hout Bay sits took 6 years to complete. Departments tend to work sequentially, and each has to complete their own processes, often including participatory processes (City official 7 2017).

(c) Restructuring

Since the early 1990s, Cape Town's administration has been in a constant state of restructuring. For the first decade after democracy this was also combined with political instability (Millstein 2010). However, since the DA has cemented its political control over the City, administrative instability has reduced. Certainly, after electoral victory in 2000 the DA ensured that it put politically sympathetic managers in place to support its service delivery plans (Cameron 2003). The administration in Cape Town also tends to have more political influence due to the adoption of the Executive Mayoral system. Introduced under the ANC, it allows the majority party or coalition to form a city government led by the Mayor, with substantial decision-making powers delegated to the Executive Mayor and the Mayoral Committee (Mayco). This system tends to encourage political power, and concomitant administrative power in the office of the Mayor, weakening the City Council and removing certain decision-making powers from the administration (Millstein 2010). Typically, however, the composition of Mayco changes more often than the management team (Bagui & Bytheway 2013).

After the local government elections in 2000, six municipalities were replaced by one metropolitan municipality. This required merging and integrating administrative functions, which has resulted in ongoing organisational challenges, fragmented operations, and struggles to co-ordinate departments and formerly territorially defined municipal responsibilities (Millstein 2010: 6). Indeed, it took one official an hour to explain to us how storm water is managed in Hout Bay and how the different departments relate to this mandate. He was located in a building 20 kilometres away from his manager and was in the process of being moved from one department to another. He explained:

> these changes are constantly ongoing, I don't know if I'm going to be staying here, I will remain with the maintenance of the river until the end of the financial year, whether there is going to be a change and whether I move to sanitation, I mean that's not for me to say, I have other matters to worry about, so I leave those with the planning people.
>
> (City official 3 2017)

A second official explained,

> I used to be part of Spatial Planning and Urban Design as a department which was part of Environment Energy and Spatial Planning. I now belong to the Transport Development Authority. ... I think, the objective is to align, you know try to get over cross-departmental obstacles, just to try and bring alignment between transport and development.
>
> (City official 7 2017)

While the intentions behind ongoing restructuring may be good, in practice it can be a barrier to development projects initiated from the community or civil society. As *in/formal south* discovered, it can be very difficult to know who to engage with and what the different departmental mandates are.

d) Spheres

The challenges of driving a developmental project relate not only to silos, mentalities, and restructuring in local government departments, but also to poor integration across the three spheres of government. In South Africa, the administrative capacity of local government was established initially in the 1996 Constitution. Here local government is clearly seen as an equal but separate sphere to national and provincial government. However, while local government is given its own level of autonomy, it still has an interdependent relationship with the other spheres. As the Premier of the Western Cape noted, 'anything you do you have to have intergovernmental relations working, otherwise one sphere of government is definitely going to kibosh something' (Zille 2017). Municipalities may have the right to govern their own affairs but they are still subject to national and provincial legislation. Thus City Hall is not autonomous; it is part of a larger system of 'cooperative governance'. National and provincial governments are called on to support and strengthen the capacity of municipalities and metros to manage their own affairs and perform their functions. The Intergovernmental Relations Act (2005) calls for co-ordination of policy and implementation of services across all three spheres (Picard & Mogale 2015: 172).

The challenge of a project such as that proposed by *in/formal south* is that in South Africa national and provincial government generally drive cross-sectoral developmental projects, or what we term 'developmental governance'. There are various reasons for this, including national government's strong belief that development management is a national project and requires centralised guidance. This, Picard & Mogale (2015: 174) argue, is a Keynesian assumption common throughout the developing world. Indeed, Scott's *Seeing Like a State* (1998) beautifully captures the desire by national governments to implement centrally managed 'developmental' projects. In South Africa the current political elite are also aware of local government being used to further apartheid spatial planning and the fact that decentralisation was strongly supported by the apartheid National Party during negotiations in the 1990s as a way to maintain meaningful control over parts of the state (Picard & Mogale 2015). Thus there are strong incentives for national government to maintain control over developmental governance.

From the perspective of civil society activists, it is not always clear which sphere of government to approach, a problem encountered by *in/formal south*. As one official explained, 'It's a myriad of competing mandated authorities. And it does become very problematic to try and get anywhere on anything' (City official 7 2017). When one senior provincial official was asked who should lead the development of the Disa River, she replied 'I don't know' and then turned to ask a junior colleague 'Whose responsibility would it be?', to which there was no forthcoming answer (Provincial officials 2015). If those working in government are unclear of the mandated boundaries there is little hope for non-state actors to access the opaque system without a political or bureaucratic champion. An expert on river management explained to us that managing the Disa River involves:

> all of national, province and city. So there's the City catchment guys … and there is the National Environmental Act that brings the national guys in and then in Province, the Department of Environmental Affairs, Development and Planning, they also have a say. So all three tend to impact on the river, maybe that's the problem that they all three pass the buck. But then when there was massive erosion in the river the other day that was affecting properties, they were all involved and then they can act quickly because it's an emergency and suddenly things get done, but for the most part it seems like things just get log jammed because it's too easy to suggest that we can't do anything because the province has to approve it and the province says that's the national responsibility.
>
> (River activist 1 2015)

e) Scale

Lastly, even if there was better co-ordination between spheres of government, or the project sat squarely at the level of local government, challenges of scale operate in the bureaucratic system. A developmental project such as the *in/formal south* one operates at the neighbourhood level, which conveniently aligns to a political ward in this case. However, City departments do not necessarily overlap with political boundaries, such as wards and subcouncils. As described in the Introduction, a senior official in the waste sector explained (City official 1 2015) that wards' and subcouncils' boundaries

> don't mean anything to us when it comes to our normal operations. They do not deliver services along an invisible political ward boundary … Because the … city is so big, we're divided into areas for management purposes … the only time we take Subcouncil or ward boundaries into consideration is when there's a ward allocation project.

The level of the 'ward' is thus a challenging scale on which to implement a developmental project, yet this is the level at which residents tend to coalesce around service delivery issues. There is thus a disjuncture between democratic and developmental boundaries and the planning associated with this.

To conclude, it is instructive to note that subsequent to the failure of the Cordaid project, *in/formal south* has moved from a non-profit approach to rehabilitation, conceptualising instead a business plan for commercialising the cleaning of rivers across South Africa. This new strategy they call 'Big Rivers'. It involves cleaning rivers with the support of corporates, in order to market water as environmentally sound, in a similar way to which coffee is marketed as 'fair trade' (*in/formal south* staff 1 2017). *In/formal south* have thus travelled full circle from a civil society-oriented developmental project to a market-focused business strategy, both of which have as their ultimate aim the cleaning of rivers in South Africa through the use of biomimicry, situated in the township and managed as part of township livelihoods.

The Hout Bay Rivers Catchment Forum (HBRCF)

Two years after the failure of *in/formal south*'s attempts to introduce a water and waste project to Hout Bay, we returned to talk to key informants involved with the river. To our surprise, while the holistic, multi-community biomimicry project was unsuccessful, there was new action from the City on cleaning up the river. In February 2016, an official responsible for River Management in Hout Bay contacted a previous member of the Hout Bay Heritage Association, whom the ward councillor had introduced him to in the past. The official asked if he would be interested in chairing a new river catchment forum for the Disa River. Catchment forums are government supported, but are civil society-driven, voluntary organisations that promote the effective management of a river and its surrounds. On his appointment, the new chair agreed to run the Forum and immediately went about linking up with the existing small, Valley-based, *Friends of the Rivers of Hout Bay*. Together with relevant City department officials they had a launch function in May 2016. While the *in/formal south* process did not directly lead to the formation of a catchment forum, it undoubtedly raised awareness of concerns with the river and indirectly triggered action from the City (River activist 3 2017; Environmental activist 2 2017).

The HBRCF has been involved, with other NGOs, in launching a series of small projects aimed at cleaning the river and dealing with waste in Imizamo Yethu and Hangberg. There are regular meetings with City and, on occasion, provincial and national departments. The HBRCF has special interest groups focusing on Water Management, Solid Waste and Biodiversity. Each of these runs their own projects and can hold independent meetings (River activist 3 2017). Projects that are underway, or earmarked for 2018, include new retention ponds, litter traps, and additional grey water disposal, all implemented by the City's Stormwater and Catchment Department. The City's Environmental Business unit has also attended HBRCF meetings to propose the *Source to Sea Initiative*, which will connect 'people, nature, businesses and organisations to the river catchment and to one another'. There is a new Imizamo Yethu waste minimisation pilot project supported by an NGO in Hout Bay and the City's Department of Solid Waste. And a weekly, 5-kilometre Park Run, drawing over 500 runners, has started along the banks of the river.

One of the most strongly supported initiatives, however, is a bio-remediation project that echoes some of the key values of the *in/formal south* project. The aim is to divert and treat heavily polluted low to medium flows with natural filtration, and work with communities in the Valley and Imizamo Yethu to do this (HBRCF 2017). The Chair of HBRCF notes that 'of everything I have tackled in the last year this is the one that has, by far, the highest level of interest ... There are many, many people who want to be involved, both from government and private' (River activist 3 2017). This indicates that *in/formal south* may certainly have played an influential role in the inspiring change and more effective management of the Disa River.

Why has the HBRCF been able to achieve at least some of what the *in/formal south* project could not? Interviews with leaders of the catchment forum and with city officials highlighted three key issues. First, having an official or a politician champion a project is essential. In the case of the Disa River, there was no politician driving a project, but when a City official saw that he could impact on the river by setting up a collaborative forum that was supported by national policy, there was movement. Previous work by *in/formal south* may have influenced this process, but it did not lead to a large-scale project as the City wants to work with those they feel are the 'right' civil society champions.

This raises the second, and critical, issue of networking and trust. The current chair of the HBRCF is well connected to both officials and politicians and has a good working relationship with both sets of actors (River activist 3; Environmental activist 2 2017; City official 3 2017; River activist 2 2017). He was able to connect with the subcouncil chair in 2016 and ensure political support for the project based on previously successful working relationships (*in/formal south* staff 1 2017). As one of the members of the forum describes, the Chair has:

> the kind of authority to call on officials to come to meetings and a manner of keeping them accountable in a very good way; he's got the city's confidence, and slowly as the meetings in Hout Bay River Catchment progressed, we've had more and more officials attending. It's been wonderful!
>
> (Environmental activist 2 2017)

The Chair himself acknowledges he has a very good relationship with the City; 'you need one person to recommend you to another, where there is mutual respect that opens the door to a conversation'. He also explains that part of the success is due to understanding the way the City works, 'You've got to work with government people to show that you're willing to complement them ... to be significantly influential, you have to understand the background and the way government officials work'. In his view all too often 'NGOs will attack government officials for not doing this, for not doing that, but actually those individuals will have no idea what's really going on behind the scenes in government' (River activist 3 2017). The importance of individual networks is echoed by a current subcouncil chair who explained that resolving a pressing service delivery issue can be 'reliant on personalities, so it's about the relationships, it's about who you can call in government to get an immediate response because they have a good relationship with you' (Political representative 2 2017).

Beyond champions and personal networks, the third reason that the HBRCF has been relatively successful is that it has not tried to launch one large holistic developmental project, but rather has taken an incremental approach to development that fits into the limited scope of various line departments. The strategic division of the Forum into special interest groups aligned to different departments has led to prompter responses from the City. It demonstrates an understanding of bureaucratic silos and mentalities as an inevitable part of governance. Allied to this approach is the view from the Chair that it is crucial to get any project into a departmental budget:

> The skill of this is to get on to any fiscus. I don't actually argue about which fiscus we get on to, which budget, any one will do. It is their job to determine overall priorities. We are representing one small portion of all the responsibilities they have. I argue our case for Hout Bay, and if they do agree it is the appropriate solution, then they will incorporate it into their planning. When I hear it is on to the fiscal planning budget I sit back with a sigh of relief, because that is what we are there to do: get it into a budget line.
>
> (River activist 3 2017)

The relative progress of the HBRCF confirms why *in/formal south* was unsuccessful. The new Forum has bureaucratic and civil society champions, with strong links to City Hall. They are able to use networks to drive development. This development, however, is divided into small projects and occurs *within* the silos of local government departments. This piecemeal tactic means they are able get into budget lines and to generate relatively speedy responses from government officials. Importantly, this approach is a reflection of successful civil society governance; it is not a reflection of a strong formal democratic system, as elected representatives are almost entirely bypassed in this politics. The HBRCF is essentially the story of connected Valley-based elites working directly 'the short route of accountability' with compatible government officials. Indeed a key objective is 'Preserving the character of Hout Bay' (HBRCF 2017). There is no attempt to challenge racial and spatial inequalities facing the ward. The HBRCF is not a reflection of wider democratic processes that take into account the overall needs of all three Hout Bay communities. This was the developmental approach *in/formal south* took, and this approach failed.

Conclusion

The *in/formal south* case demonstrates some of the challenges facing communities and civil society in successfully initiating development projects through the institutions of local democracy. Key here are three sets of factors: first, the *in/formal south* development project enjoyed significant support across the divided communities of Hout Bay, and from officials in diverse departments in the City, but was unable to present itself to City Hall in a way that addressed key features of bureaucratic governance. This foregrounds the reality that the bureaucracy consists of multiple departments divided into silos with different mentalities; ongoing and destabilising internal restructuring; and misaligned scales of bureaucratic and democratic governance. Consequently, the *in/formal south* project ended up a democratic round peg unable to fit into a bureaucratic square hole.

Second, and relatedly, the *in/formal south* project failed to win support from senior political levels of the subcouncil and City that may have enabled project implementation across line departments. Third, even if the project did secure the support of the Mayor, it would still have required co-operation with other spheres of government, including the national level, which is controlled by the ANC, the main rival of the leading party in the City of Cape Town. All these factors add layers of institutional and political complexity to a development project in a local area that is close to insurmountable for civil society-led initiatives without major political or economic backing. Thus, while local democratic institutions can have some influence over City Hall (if they can target the Key Performance Areas of line departments), they have none over provincial and national spheres.

These levels of democratic disconnect in the *in/formal south* case are confirmed in the negative by the more effective approach of the Hout Bay River Catchment Forum (HBRCF), an invited space set up by City Hall but occupied by river activists networked into the local state. Drawing on the social capital of long-standing relations of trust with individuals in the City, these actors know how to work within the logic of the City's line departments rather than across them. They also embrace a more incremental, piecemeal and modest set of goals largely within the power of the local state, and without extensive democratic consultation or attempts at popular coalition-building across the diverse communities of Hout Bay. Consequently, an effective river catchment forum can generate ad hoc projects, but it is unlikely to tackle more systemic issues of racial inequality, land use and disparate health concerns facing township residents. In short, for all its relative success in delivering a cleaner river, the strategy of working 'within the bureaucracy' is disconnected from democracy as a process of building the demos.

Note

1 An organism used in scientific testing to determine the level of faecal contamination in a body of water.

References

Actuality. (2017). *Water Integrated Catchment Management.* www.acturban.com/hout-bay-water-integrated-catchment-management/. Accessed 10 September 2017.

Anciano, F. (2018). Clientelism as Civil Society? Unpacking the Relationship between Clientelism and Democracy at the Local Level in South Africa. *Journal of Asian and African Studies, 53*(4), 593–611.

Bagui, L., & Bytheway, A. (2013). Exploring E-Participation in the City of Cape Town. *The Journal of Community Informatics, 9*(4). http://ci-journal.org/index.php/ciej/article/view/982. Accessed 8 May 2018.

Bardouleau, M. (2010). *Hout Bay Disa River Pollution Protest.* www.savingwater.co.za/2010/05/31/13/hout-bay-disa-river-pollution-protest/. Accessed 10 September 2017.

Barichievy, K., Piper, L., & Parker, B. (2005). Assessing 'Participatory Governance' in Local Government: A Case-Study of Two South African Cities. *Politeia, 24*(3), 370–393.

Beresford, A. (2015). Power, Patronage, and Gatekeeper Politics in South Africa. *African Affairs, 114*(455), 226–248.

Biomimicry Institute. (2017). *What is Biomimicry?* https://biomimicry.org/what-is-biomimicry/. Accessed 18 September 2017.

Cameron, R. (2003). Politics–Administration Interface: The Case of the City of Cape Town. *International Review of Administrative Sciences, 69*(1), 51–66.

CTP (Cape Town Partnership). (2017). *About Cape Town Partnership.* www.capetownpartnership.co.za/about/. Accessed 3 May 2017.

CCT (City of Cape Town). (n.d.). *Departments.* www.capetown.gov.za/Departments. Accessed 14 November 2017.

CCT. (2003). *Subcouncil's System of Delegations.* http://resource.capetown.gov.za/documentcentre/Documents/Procedures,%20guidelines%20and%20regulations/System%20of%20Delegations%20for%20Subcouncils.pdf. Accessed 8 October 2017

Cohen, J., & Fung, A. (2004). Radical Democracy. *Swiss Journal of Political Science, 10*(4), 23–34.

Cronje, J. (2014). E Coli Alert for Cape beach. *IOL online,* 21 September. www.iol.co.za/news/south-africa/western-cape/e-coli-alert-for-cape-beach-1754022. Accessed 10 September 2017.

Dawson, H. J. (2014). Patronage from Below: Political Unrest in an Informal Settlement in South Africa. *African Affairs, 113*(453), 518–539.

Froestad, J. (2005). Environmental Health Problems in Hout Bay: The Challenge of Generalising Trust. *South Africa Journal of Southern African Studies, 31*(2), 333–256.

Froestad, J., Grimwood, S., Herbstein, T., & Shearing, C. (2015). Policy Design and Nodal Governance: A Comparative Analysis of Determinants of Environmental Policy Change in a South African City. *Journal of Comparative Policy Analysis: Research and Practice, 17*(2), 174–191.

HBRCF. (2017). Draft Minutes Hout Bay Rivers Catchment Forum. 7 September 2017.

HBRRA (Hout Bay Residents and Ratepayers Association). (n.d.). Article 1: Addressing the issues of Imizamo Yethu and Dontse Yakhe: the Residents' Association takes action: Extracts from Environmental Impact report on Dontse Yakhe by independent Environmentalist Consultant Andre van der Spuy. www.houtbay.org.za/SupplementHoutAbout201006.html#IYissues. Accessed 10 September 2017.

in/formal south. (2015). Notes on Biodiversity in Hout Bay. *In/formal south.*

Millstein, M. (2010). Limits to Local Democracy: The Politics of Urban Governance Transformations in Cape Town. *Working Paper 2.* Visby: Swedish International Centre for Local Democracy (ICLD). https://icld.se/static/files/forskningspublikationer/icld-wp2-printerfriendly.pdf. Accessed 16 May 2018.

Picard, L. A., & Mogale, T. (2015). *The Limits of Democratic Governance in South Africa.* Boulder, CO: Lynne Rienner Publishers.

Piper, L., & Deacon, R. (2008). Party Politics, Elite Accountability and Public Participation: Ward Committee Politics in the Msunduzi Municipality. *Transformation: Critical Perspectives on Southern Africa, 66*(1), 61–82.

Piper, L. Anciano, F. Sikota, Z. Williams, R., & Muteti, A. (2015). Water and Waste in Hout Bay: Stakeholder Mapping for Cordaid. Sustainable Livelihoods Foundation Report.

Scott, J. C. (1998). *Seeing like a State: How Certain Schemes to Improve the Human Condition have Failed.* New Haven, CT: Yale University Press.

Thompson, L. (2014). Agency and Action: Perceptions of Governance and Service Delivery Among the Urban Poor in Cape Town. *Politikon, 41*(1), 39–58.

World Bank. (2004). World Development Report: Making Services Work for Poor People. *The International Bank for Reconstruction and Development.* https://openknowledge.worldbank.org/bitstream/handle/10986/5986/WDR%202004%20-%20English.pdf?sequence=1. Accessed October 2016.

Interviews

1. City official 1. (2015). Interviewed by Laurence Piper. 7 April 2015.
2. City official 3. (2017). Interviewed by Fiona Anciano. 13 September 2017.
3. City official 4. (2015). Interviewed by Fiona Anciano. 20 March 2015.
4. City official 5. (2015). Interviewed by Fiona Anciano. 20 March 2015.
5. City official 6. (2015). Interviewed by Fiona Anciano. 20 March 2015.
6. City official 7. (2017). Interviewed by Fiona Anciano. 29 February 2017.
7. City official 8. (2015). Interviewed by Fiona Anciano. 27 March 2015.
8. Community leader 1. (2015). Interviewed by Laurence Piper. 24 February 2015.
9. Environmental activist 2. (2017). Interviewed by Fiona Anciano. 26 October 2017.
10. HBP staff. (2015). Interviewed by Fiona Anciano and Laurence Piper. 30 January 2015.
11. Health Forum leader. (2015). Interviewed by Laurence Piper. 10 March 2015.
12. *in/formal south* staff 1. (2017). Interviewed by Fiona Anciano. 22 February 2017.
13. *in/formal south* staff 2. (2017). Interviewed by Fiona Anciano. 14 September 2017.
14. Political representative 1. (2017). Interviewed by Fiona Anciano. 21 February 2017.
15. Political representative 2. (2017). Interviewed by Fiona Anciano. 18 October 2017.
16. Provincial officials. (2015). Interviewed by Fiona Anciano and Laurence Piper. 24 March 2015.
17. Residents. (2012). Workshop on insecurity in Imizamo Yethu with foreign residents. 18 May 2012.
18. River activist 1. (2015). Interviewed by Laurence Piper. 16 April 2015.

19. River activist 2. (2017). Interviewed by Fiona Anciano. 12 September 2017.
20. River activist 3. (2017). Interviewed by Fiona Anciano. 15 September 2017.
21. Waste activist. (2015). Interviewed by Fiona Anciano and Laurence Piper. 20 February 2015.
22. Zille, H. (2017). Premier of the Western Cape. Interviewed by Laurence Piper. 26 September 2017.

Observation

1. *in/formal south* project meeting. 2015. Observed by Fiona Anciano. 29 July 2015.
2. City meeting (2015). Observed by Fiona Anciano. 20 March 2015.

3   Selling the mountain

Property, housing, and neo-apartheid segregation

In July 2009, an article in the *Sunday Times*, the leading national newspaper, claimed that Sentinel Mountain in Hout Bay was on the property market by auction, and that enquiries had been made by 'talk show host Oprah Winfrey, hotel magnate Sol Kerzner, Donald Trump jnr as well as the Bill and Melinda Gates Foundation'. In addition, the article continued, 'the new owners could, if they wished … name the peak after themselves as the Sentinel was not a registered trademark. Auctioneers had reportedly turned down two offers, including one for R15 million' (News24 2009).

Not surprisingly, alarm spread through Hangberg, the major settlement on Sentinel Mountain. On 16 July 2009, the morning of the auction, a crowd of 300 protesters from Hangberg gathered outside the site of the sale, the Chapman's Peak Hotel, led by the Hout Bay Civic. According to Isaac James, a Hout Bay Civic leader, they wanted to 'sit down with the auctioneers' to convince the owners to halt plans to sell the prized real estate (News24 2009). However, once it was clear that the auction was proceeding, the protest became confrontational. Some protesters began to throw stones, and the police opened fire on the crowd, showering them with rubber bullets and teargas. This confrontation quickly brought the auction to a halt.

In the aftermath of this event, it transpired that it was not the whole of Sentinel Mountain that was for sale, which would have included Hangberg on its lower reaches. Instead, it was two portions above the settlement, officially Erf 3557, bordered on either side by the Table Mountain National Park. The land was owned by G&R Marine Services CC that bought it in 2003 for R60,000. The lower portion of 60 hectares abutting Hangberg was sold off to SANParks for R800,000 shortly after the protest (Moneyweb 2010). The higher portion of 100 hectares that remained was surrounded on every side by the national park, and negotiations ensued to sell it to SANParks.

The *Sentinel* protest shines a spotlight on the precarious land rights of long-standing residents in Hangberg, many of whom had suffered displacement under apartheid, and the majority of whom do not have private tenure rights nor the wealth to acquire them. Despite living for generations on Sentinel Mountain and working on the sea of Hout Bay, Hangberg

Figure 3.1 The global property market comes to the fishing village

residents' right to belong appeared to be under threat, this time by the (global) property market (see Figure 3.1). While this particular case was driven by a misperception fanned by the media, it heightened insecurities of belonging in Hangberg significantly and, according to some local leaders, contributed substantially to the 'Battle of Hangberg' in 2010 (PMF 2017), that is, the violent resistance that followed attempts to remove informal settlers discussed in Chapter 4.

In addition to insight into the politics of place in Hout Bay, the incident of the 'sale of the Sentinel' is an entry point to the political economy of urban property in post-apartheid South Africa, and indeed in many cities in the world where market relations shape who gets access to what kind of place. In particular, it alerts us to the impact of international and national market relations on urban land, and the way that market actors and especially property developers can shape middle-class and wealthy urbanisms, but in ways that often impact directly and indirectly on the poor. Indeed, this outcome is almost assured given the weak spatial governance of urbanisation in South Africa, and the demonstrable incapacity of the state to supply enough housing for working and poor in-migrants. Thus, *where* people live in the city is still a product of processes beyond immediate state control, and therefore disconnected from formal democracy.

In addition to accounting for where people live in Hout Bay, this chapter explores some dimensions of *how* they live too, by tracing the relationship between market, developmental, and informal governance and the urbanisms of local place. By urbanisms, we mean both the nature of the built environment, its spatial layout and architecture, and the everyday needs that living in this built environment imposes. We begin by outlining the agenda-setting power of market forces in shaping the spatiality, design and lifestyles of middle-class and wealthy residents in Hout Bay. Unable and perhaps reluctant to change fundamental patterns of land occupation, the state focuses its transformatory energies, and significant resources, on developmental governance such as housing in poor areas like Imizamo Yethu and Hangberg. At the same time, the state is incapable of keeping up with the demand for housing created by urbanisation and, in this gap, informal settlement and practices have grown.

The combined effects of market, public, and informal governance around residential property in Hout Bay is dividing residential property into 'green' areas, 'post-public' wealthy places governed by market relations, and 'grey areas' of blurred formality/informality where residents are treated by the state as populations to be governed and/or informal encroachers who attempt to avoid the state. We term this emergent spatial segregation of settlement neo-apartheid as it has both social (racial) and political (divergent forms of governance) features similar to the past, but also differences – it is driven by economic processes, and the developmental energies and financial investment of the state is now focused mostly on poor, black areas.

In outlining the evolution of neo-apartheid, we demonstrate the substantial 'power to produce' the built environment of both market actors and residents themselves compared to state actors. Indeed, both often proceed against the will of some set of local residents, including long-standing white residents of the Valley. This relative power of market and informal practice, combined with the City's de facto lack of authority over spatial development, exposes a disconnect between local democracy and attempts to manage urbanisation in Hout Bay.

The Valley: from white countryside to wealthy security estates

Chapter 1 described the evolution of Hout Bay from an agricultural and fishing area to a white residential area, especially from the 1950s onwards. To a significant extent the transformation of farming land into residential land in the Valley followed the pattern of the 'crabgrass frontier' that Kenneth Jackson (1985) coined to describe the large plots, with manicured lawns and leafy gardens, that typified the growth of suburbia in the United States. From the 1980s, however, and especially after apartheid, the pattern of settlement began to change with the densification of the suburbs, the emergence of new malls, and the growth of security estates.

Notably this kind of transformation at this time was not unique in South Africa, nor globally. Martin Murray (2015: 180) argues that many cities around the world began to produce a post-suburban landscape from the 1970s and 1980s as 'edge' cities formed when residential suburbs evolved into 'semi-autonomous' outer cities, situated along orbital roads, 'that rival and often surpass the traditional big city downtowns as vibrant centres of socioeconomic power and commercial vitality'. Murray places private real estate developers at the centre of this transformation. Oriented to affluent consumers, they provide residential neighbourhoods 'with varying degrees of exclusivity, shopping arcades and recreational facilities' (Murray 2015: 183). Hence the emergence of wealthy 'enclaves, islands and bubbles' that are 'post-public spaces' of shopping malls, themed casinos, redeveloped urban commercial zones and themed entertainment centres (ibid. 184). In this consumerist landscape, shopping malls have become the informal capitals of suburbia: 'they represent places to congregate, to socialise, and to seek ambience and connections with others' (ibid. 186).

While not an 'edge city' as it is too small and lacks enough of the commercial and light industrial aspect, Hout Bay reflects many of Murray's features of a developer-driven landscape. This includes the growth of an Olympic size mall in the central shopping area, the dramatic growth in property prices in the Valley, and especially the flourishing of a significant number of gated communities, particularly in the last 10 years. Consequently, private property today in the Valley is expensive. According to *Property24* (2016) the average sale price for houses in Hout Bay in 2016 is R3,600,000 (US$300,000), more than double the boom price of 10 years previously. Further, in an interview with an online property website, an estate agent from Lew Geffen Sotherby's International Realty stated:

> In 2011 the average house price was R2.81 million and sectional title properties sold for an average of R1.2m … By 2015 the average house price had almost doubled to just under R4.1m and the sectional title median price had increased by 41% to just over R1.7m.
> (Terri Steyn, Lew Geffen Sotherby's, Propertywheel 2016)

She added that there has also been a marked drop in the length of time properties remain on the market:

> In 2011 the average length of time across the board was 159 days, with several properties remaining on the market for more than 400 days and some even for longer than 500 days … In 2015 the average time spent on the market had dropped by 21% to 126 days with very few properties remaining on the market for longer than 200 days, but desirable properties in secure estates sell much faster.
> (Terri Steyn, Lew Geffen Sotherby's, Propertywheel 2016)

Another estate agent gave a similar average house price of R3.5 million. He also noted that Hout Bay features in the top five suburbs in Cape Town in

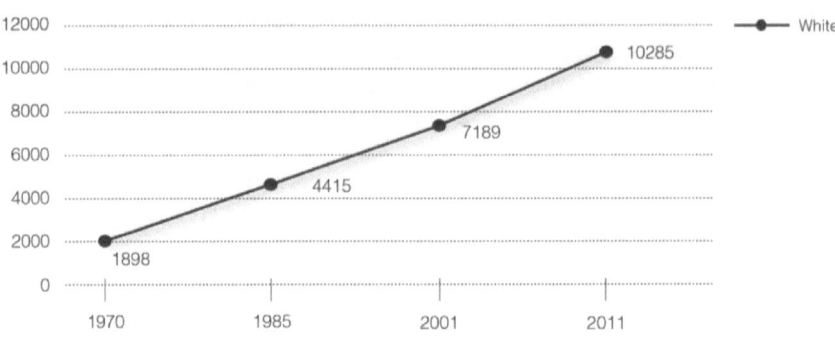

Figure 3.2 White population Hout Bay, Census 1970–2011

terms of growth price over the last 5 years, and is in the top ten most expensive suburbs overall (Estate agent 1 2017), in a city that houses nine of the ten most expensive suburbs in South Africa (BusinessTech 2017).

According to these estate agents, the reason for this increase in prices is a combination of a quick, congestion free drive to the CBD, and a wide range of convenient amenities including Woolworths, QuikSpar, Pick n Pay and Checkers, four pre-primary schools, five primary schools and three high schools, as well as the Hout Bay International School for all levels of education (Estate agent 1 2015, 2017). The sustainability of the International School confirms that there are significant numbers of wealthy foreign residents who live in Hout Bay, or at least people with money and ambitions for a cosmopolitan lifestyle. Thus, not only is there high migration into Hout Bay, but, as revealed in Figure 3.2, there has been high migration of white and wealthy people too. Last, but not least, Hout Bay is also a popular holiday destination with one respondent estimating that '20% of the fancy houses are holiday homes' (Environmental activist 1 2017).

Central to the growth of private real estate in Hout Bay is the emergence of security estates, especially in the last 10 years. As revealed in Table 3.1, Hout Bay has at least nine major security estates occupying around 14 per cent of the total residential space. On a reasonably accurate estimate drawn from the databases available to estate agents, they are worth R5 billion in 2017 prices (Estate agent 2 2018). According to *Property24* (2016), there are 2308 estate houses in Hout Bay, 3469 houses and 972 apartments. Thus, security estate houses constitute 35 per cent of all formal residential structures. Assuming that the lower estimate of *Property24* of an average value of R3 million/house and R1.45 million per apartment is correct, this puts the total value of formal residential housing in Hout Bay at R19 billion. This excludes commercial and industrial property. By way of comparison, the City of Cape Town's annual capital budget is around R6.8 billion and operating budget is around R37.5 billion (CCT 2017).

Table 3.1 Major security estates in Hout Bay

	Name	Hectares	Median valuation 2017	Number of Erven	Total (R million)
1	Avignon Estate	15	R5.8 million	67	388.60
2	Beach Club Estate	3.06	R4.1 million	64	262.40
3	Bergendal Estate	18.2	R5.95 million	133	791.35
4	Hanging Meadows Estate	1.8	R7.75 million	27	209.25
5	Kenrock Estate	34	R7.7 million	108	831.60
6	North Oaks Estate	3	R5.85 million	41	239.85
7	Ruyteplaats Estate	37.2	R8.1 million	198	1603.80
8	Stoneybrook Estate	6.5	R4.55 million	12	54.60
9	Tierboskloof Estate	18	R7.25 million	81	587.25
Total		**137/953 14%**	**R6.8 million**	**731**	**R4.9 billion**

Importantly, the advent of gated communities adds another important dimension to governance in Hout Bay. As Ballard and Jones (2015: 298) observe, a key reason for their popularity among middle-class and wealthy residents is the well-ordered, apparently consensual (but often draconian) governance of these places. Thus, in addition to the rowdy politics 'outside the walls' where the HBRRA champions the interests of more conservative members of the valley, whether in respect of the toll road (Chapter 8) or the expansion and development of Imizamo Yethu (Chapter 6), we also have politics 'inside the walls' of the gated community. This system of micro-managing lawns, house frontages, and noise levels has a security component that we explore further in Chapter 9.

As already suggested, the dramatic growth in private property in Hout Bay, exemplified by the phenomenon of security estates, is not unique to Cape Town and is a global phenomenon. Thus Murray's account of the emergence of 'edge cities' noted above resonates with Harvey's (2012) argument, inspired by Lefebvre (1991), that urbanisation allows for new forms of capital accumulation. On this view, surplus capital is invested in new urban projects that, in good times at least, yield excellent returns. Ballard and Jones (2015: 298) offer a good example of this in the transformation of sugar farmland into security estates, thereby radically increasing the value of space in the process. However, there is more going on than developers making money out of urbanisation, or the commercialisation of space rather than of things. Again following Lefebvre, Harvey (2015: 9) points out how suburbanisation was also about constructing a particular lifestyle to consume the products of manufacturing capitalism (the car, fridge, and air conditioner), as the new wealthy enclaves come with a commodification of lifestyle for the middle classes. This, Sharon Zukin terms, 'pacification by cappuccino' (in Harvey 2012: 14):

The postmodernist penchant for encouraging the formation of market niches, both in urban lifestyle choices and in consumer habits, and cultural forms, surrounds the contemporary urban experience with an aura of freedom of choice in the market, provided you have the money and can protect yourself from the privatization of wealth redistribution through burgeoning criminal activity and predatory fraudulent practices (which have everywhere escalated).

In the South African context, these dynamics of market governance in constructing the urban landscape and lifestyles overlay a history of racial conflict that gives crime an identity dimension too. Indeed, fear of crime is a major rationale driving the rise of the 'gated community' or security estate in South Africa. As Ballard and Jones (2015: 298) observe, for wealthy white people, gated communities represented a seemingly well-ordered and consensual alternative to 'the perceived mayhem of local politics outside the walls'. From the outside, however, gated communities appear as the symbol of inequality and lack of racial integration in South Africa, leading the courts, the South African Human Rights Commission, and senior politicians to question their virtues. Some have even termed them a 'new apartheid' (Lemanski 2004: 101).

Lastly, the power of property developers and wealthy individuals to shape the post-apartheid city may not be unusual, globally speaking, but it is surprising in South Africa given the express commitment to creating opportunities for historically marginalised groups to access urban land (see the Development Facilitation Act (DFA) of 1995, and more recently the Spatial Planning and Land Use Management Act (SPLUMA) of 2013). Ironically, as Ballard and Jones (2015) observe, the DFA has equally been used by developers to transform agricultural land into gated communities. Indeed, a number of observers comment on how the state seems unable to rein-in private developers. Murray (2015: 186) writes that:

> Starting in the 1990s and continuing for at least a decade, the combination of weak planning frameworks, lax regulatory mechanisms and inconsistent code enforcement provided real estate developers with a great deal of discretion to build without much oversight throughout the property boom of the early 2000s.

Similar points are made by Lemanski, Landman, and Durington (2008), and Ballard and Jones (2015: 303) who note that a key constraint on local authorities is that many planning applications are heard at the provincial level, and national legislation allows for fast-tracking of many developments, leaving the provinces little control. Indeed, local governments will often remove objections to a development if the developer assists with local infrastructure, for example building a road or ramp to the freeway. These investments are beyond the boundaries of the gated community, but help clear the way for permission from local government. Ultimately, Ballard and Jones conclude that there is little that local authorities can do to stop private developments 'other than to ensure that they are not saddled with the costs of providing infrastructure' (2015: 309). Consequently, to make cities more equitable, local authorities focus their efforts on the poorer communities.

Obviously in the case of Cape Town, similar constraints on the role of local government in controlling planning applications apply as in other cities of South Africa, but there is the additional consideration that, as a city run by the self-proclaimed 'business-friendly' Democratic Alliance, it actively encourages market development. In this respect Huchzermeyer (2011: 47–52) notes that Cape Town has embraced the global branding of 'competitive cities' that looks to make the city attractive to 'hyper-mobile capital', both economic and human. These notions, central to the ideal of 'world class cities' are evident in the various vision documents embraced by Cape Town, and how they have been implemented around public–private partnerships.

Evidence for this includes the City's 2016/2017 Integrated Development Plan (IDP) that lists as the first point under its threefold vision: 'To be an opportunity city that creates an enabling environment for economic growth and job creation, and to provide help to those who need it most'. The discourse of creating 'opportunity for growth' where the role of the state is to 'enable' this, resonates with neo-liberal conceptions of the role of the market and state respectively (CCT 2016). In addition, there are initiatives from the business community such as 'Accelerate Cape Town', that claim success is making Cape Town a more business-friendly city 'like Barcelona or Dubai'. Aiming to build alliances between business, government, and academia, Accelerate Cape Town hosts networking events that have been addressed by people like Boris Johnson and Helen Zille. Indeed, they claim success in attracting new business to Cape Town. Thus,

> both Philip Morris and Johnson & Johnson have made strategic decisions not to leave or close their Cape Town operations ... Steinhoff have moved their head office to the Cape and DHL Express have established their sub-Saharan head-offices in Cape Town.
>
> (Swinhoe 2012: n.p.)

This approach reflects a desire to build the kind of government–business alliance that inspired Clarence Stone's urban regime theory, and the DA government appears to like this. Thus, in August 2011, the Western Cape Minister of Economic Development, Mr Alan Winde, stated that 'if we are to succeed in growing business and attracting investment, we must eliminate red tape and roll out the red carpet'. Hence, he announced a programme 'to eradicate red tape and create an enabling environment so that local businesses may grow and employ more people, and so that we may become an attractive destination for foreign companies to invest in' (Winde 2011).

However, even assuming a government–business regime in place in DA-run spheres of the state, more is required to demonstrate that the City of Cape

Town places the interests of investors ahead of local residents' issues. In this regard, a number of articles have appeared in the media complaining about the DA government's cosiness with private developers, as summarised in the 'from red tape to red carpet' approach that has seen 'developmental battles' emerging in the Bo-Kaap, Constantia, and Maiden's Cove (Johnson 2016; see also Ashton 2013). Indeed, very similar arguments have been made of Hout Bay by Len Swimmer, Chair of the Hout Bay Residents and Ratepayers Association (HBRRA), complaining both of the lack of public participation around key aspects of urban planning, and even an attempt in 2013 to repeal the City's Public Engagement Policy (Swimmer 2013a). Writing in the HBRRA newsletter of the same month, Swimmer (2013b) reported HBRRA objections to 'opportunistic sub-division applications' in the Valley, stating:

> The basic idea behind the desire to reduce bureaucracy in land use planning is admirable in itself. Unfortunately, corners tend to get cut in the process. You may be aware of the considerable exposure in the local press recently of the attempt by the mayor and her officials to alter arrangements so that subcouncils will no longer have the power to decide land use planning matters concerning their wards. This move will stop pesky junior councillors in subcouncils from making decisions that go against the top politicians' desire to turn Cape Town into a wall-to-wall concrete jungle and remove entirely the public's ability, as Interested and Affected Persons, to influence the final decision on any such matter.

In this regard, it is notable that, according to a former city official, attempts by the city to support densification or 'gentrification' of Cape Town were often met with resistance by Subcouncil 16, where politicians from Hout Bay acted to defend the 'rural character' of the area (Denoon-Stevens 2017). Other reasons given for refusing new subdivisions and development included the traffic impact and harm to tourism potential, but mostly it was the threat to 'the rural character of Hout Bay' that formed the basis of objections. Usually connected to the HBRRA, these politicians tended to defend the interests of the organised white and wealthy residents only (Provincial officials 2015; Air pollution activist 2017). As Denoon-Stevens notes (2017), these objections were usually overturned by city or provincial appeals committees that put more emphasis on densification and other policy considerations.

Notably, despite the efforts of Swimmer and the HBRRA, in 2015 the City passed a municipal planning by-law that allows both smaller plots and proportionally more floor space per plot to facilitate densification (CCT 2015). According to an estate agent this will make it easier to subdivide and add a second dwelling on existing plots for around 7,000 properties that now qualify under the new by-law. This creates new incentives to maximise land value and thus 'developers are searching for opportunities and by my count there are 200 new properties coming onto the market in the next two years alone' (Estate agent 1 2017).

While the actions of politicians in Subcouncil 16 can be seen as an attempt by representatives of Hout Bay to defend local interests (albeit of a segment of high-income residents), it is noteworthy how these served to delay rather than prevent the implementation of the city policy of densification. Further, Denoon-Stevens holds that with the advent of SPLUMA in 2013 and the decision to establish a Municipal Planning Tribunal, the City of Cape Town has even more capacity than previously to impose a central vision on the suburbs and settlements of Cape Town than before. In short, local influence over market development in Hout Bay is thin to non-existent. Not only do individuals or companies, rather than residents collectively, decide what to buy or sell in terms of state-backed property relations, but even the influence over key property rules is limited. For example, the minimum size of properties is defined at the level of the city, not the suburb. Furthermore, these framings are informed by the interests of the city as a whole, which, under the DA, have a neo-liberal frame, rather than in terms of the preferences of Hout Bay residents (or those who succeed in speaking in the name of Hout Bay residents).

Imizamo Yethu: developmental governance, informality, and the limits of state housing

The story of housing in the valley is one of market development driving formal settlement, at remarkably profitable rates, whereas in the poor settlement of Imizamo Yethu there is no formal market development. Rather, formal housing is driven by the state, in some cases assisted by international donors like the Niall Mellon Foundation, as discussed further in Chapter 6. Indeed, as the City of Cape Town frequently states, a large proportion of its budget is spent on townships and informal settlements annually.

Although the DA claimed in its 2016 local government election manifesto that 67 per cent of the City budget is spent in poor areas, fact-checking of this claim put the real figure somewhat lower. The exact amount is hard to identify as budget allocations are done by function to line departments and include city-wide costs that all benefit from, which makes an accurate calculation difficult. According to AfricaCheck (2016), the real figure is probably less than 50 per cent. This notwithstanding, there is no doubt that the City spends much more on the urban poor today than it did under apartheid. Furthermore, there are additional funders of developmental governance in townships, including donors and other spheres of government. When it comes to housing, provincial government is an especially important player; there is no doubt that the state is the main source of investment in public housing in Hout Bay and that this has occurred exclusively to date in the poor, black settlements of Hangberg and Imizamo Yethu.

Notably, as all formal housing in Imizamo Yethu is supplied through state-sanctioned schemes, it cannot be sold legally, or rented out, for 8 years (Section 10a (1)-(4) HAA 2001). Consequently, banks will not accept government housing as collateral on loans. Rather than a means to give poor people access to the formal housing market as individual property rights-bearers, government housing serves more the developmental function of sheltering the poor. In this way, we argue, the state housing schemes treat the poor as populations to be managed by the state rather than as legal rights-bearers in the market place.

A common argument is that developmental governance usually has a democratic character in that recipients are involved in project implementation, often electing a management committee for example. While different government programmes have different models that involve various degrees and kinds of resident involvement – we will discuss the history of housing projects in more detail in Chapter 6 – no national policy framework invites recipients to frame policy in substantive ways relevant to local conditions. Rather, housing lists are drawn up and allocated by government, ostensibly on the basis of 'need'. But the actual process tends to be a lot more political and sometimes corrupt (SERI 2013), and in the case of Imizamo Yethu, usually mediated through party-political networks. Occasionally there are instances, such as arguably with the Niall Mellon project in Imizamo Yethu in 2002 where local leaders played a role in securing a housing project for the settlement, but most of the time popular influence tends to be around the details of implementation. When this occurs in a fair and transparent way, developmental governance can be said to be implemented democratically. Other than this, there is no collective control by local residents over who gets what kind of state housing, and when they get it.

State provision of housing has not been able to keep up with popular demand in Imizamo Yethu. Thus, over a 30-year period the population of Imizamo Yethu has grown from 1,800 to 8,063 to 15,538 in 2001, but formal housing has grown to just 1,393 units and informal to a massive 4,613 in 2011. Clearly, the attempt to meet the need for shelter through state-sanctioned developmental governance is not keeping pace with urbanisation. There is consequently a massive informal market in housing in Imizamo Yethu, like any other poor settlement in South Africa. This takes the form of the informal sale of houses, although this is very rare, but mostly of rental of backyard space, rooms in the house, or of shacks. Indeed, the price of informal housing is also affected by limited supply of housing more generally in Hout Bay, with an entry level shack in Imizamo Yethu selling for around R7,500 in 2012 (Community leader 2 2012), and today up to R15,000. Renting a single room in the most modest shack costs R500 per month, with an additional R200 per month for electricity (Community leader 2 2017; Trader 2017).

Thus, by 2011 just 23 per cent of all housing in Imizamo Yethu was formal. In Chapter 1 we describe the changing profile of immigrants into Imizamo Yethu, noting that in the last 10 years a significant proportion have been migrants from the rest of Africa, and constitute about 15 per cent of the settlement. However, even excluding foreign migrants, state supply is way below popular demand. Importantly, as revealed in Figure 3.3 this does not

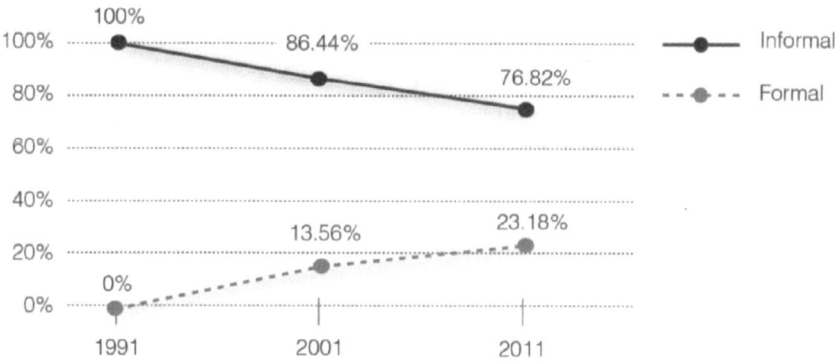

Figure 3.3 Percentage housing type in Imizamo Yethu 1991–2011

mean that the state is doing nothing; indeed, the proportion of formal versus informal dwellings in Imizamo Yethu is growing down time, but just not at a rate that can match urbanisation. Once again, the state is simply too slow and lacks the resources to keep up with either market actors or poor migrants.

In practice then, informality has grown in Hout Bay since the demise of apartheid. Cape Town has seen a significant growth in informal settlements in the last 20 years, with an estimated 376 in 2015, amounting to around 260,000 structures or between 500,000 and 750,000 people (City official 1 2015). In the case of Imizamo Yethu, the initial objective of the settlement in 1991 was to provide formal plots and then housing for those in the five informal settlements around Hout Bay. To this end, 18 hectares of state land were set aside for housing for 455 families, and a further 16 hectares for supporting facilities. The key assumption, at least on the side of the local government and residents of the Valley, was that migration into Hout Bay would not increase, but nothing could be further from the truth. Indeed, if anything, the establishment of a formal settlement attracted more rather than fewer migrants. This has given rise to a long-standing and bitter feud over what to do with the 16 hectares of land in Imizamo Yethu discussed in Chapter 6.

The inability of the state to provide housing to match demand from poor migrants means that the formal settlements for poor people quickly become what Yiftachel (2009) terms 'grey spaces', between the 'whiteness' of formality and legality, and the 'blackness' of informality and illegality. In particular, they become messy zones of formal and informal housing, as well as formal and informal livelihoods. Emblematic of this blurring into greyness in respect of housing is the growth of backyard shacks. This involves the rental of space behind a formal house to someone who builds his or her own shack, and the informal connection to utilities, such as electricity, supplied to the formal house. As noted, 'backyarding' is very common in Imizamo Yethu as it is in most townships in South Africa. Invariably then, the growth of informal settlements happens adjacent to formal areas, as is the case in Imizamo Yethu, with the main informal area, Dontse Yakhe, stretching up the Skoorsteenkop mountain.

The urbanism of residents of grey spaces has both negative and positive aspects. The growth of informal accommodation in and around formal settlements contributes greatly to their densification, including the infilling of open land which usually means the loss of public open public. This is a key issue in Imizamo Yethu, for instance, where there is no public space other than the street or the shebeen. In addition, densification places significant strain on infrastructure such as electricity, roads, and especially sanitation that are all designed to supply a lower demand. The impact of large numbers of poor people living in informal conditions on health, security, and the environment are largely negative and well documented in the case of Imizamo Yethu (Harte, Sowman, Hastings, & Childs 2015). Living in the grey space of a poor settlement in South Africa comes with many risks additional to living in a small and often poorly built structure.

There are, however, also many positive elements to urbanisms emergent in grey spaces too. Across the world informal settlements are sites of the construction of new urban identities that mediate national, racial, and ethnic differences; they are sites of labour and energy to drive emerging growth in the city; and are even sites of political innovation, mostly famously argued by Holston's (2008) account of insurgent citizenship. Finally, yet importantly, are the forms of livelihoods practices that unemployed and underemployed people must embrace to survive, often leading to innovations that some, like de Soto (2002), see as potentially driving future economic growth. Indeed, in Imizamo Yethu there is significant evidence of informal economic activity. As reflected in Figure 3.4, a business census conducted by the Sustainable Livelihoods Foundation in 2013 found no less than 600 micro businesses, almost all of which are informal.[1]

What this SLF survey reveals is a great deal of economic activity, with one business for every 25 residents, or one for every 10 households. It also reveals the 'greyness' of this activity for, while most of it is retailing formal products, some of it is clearly illegal, such as the more common enterprise of 182 shebeens or liquor retailers in Imizamo Yethu. Lastly, while liquor retail is dominated by South African shopkeepers, it is notable that spaza'shops (local grocery and convenience stores) are mostly run by immigrant groups, especially from East Africa. The preponderance of immigrant entrepreneurs in parts of the informal sector, most of whom are unable to find legal employment, helps explain some of the community dynamics that sometimes result in xenophobic attacks in townships around South Africa (Charman & Piper 2012).

Perhaps the job of supplying houses for the urbanising poor was always going to be an impossible one for a middle-income country like South Africa, but there are clear policy choices that exacerbate the problem. As Huchzermeyer (2011: 53) notes, in following the global trend of branding Cape Town as a

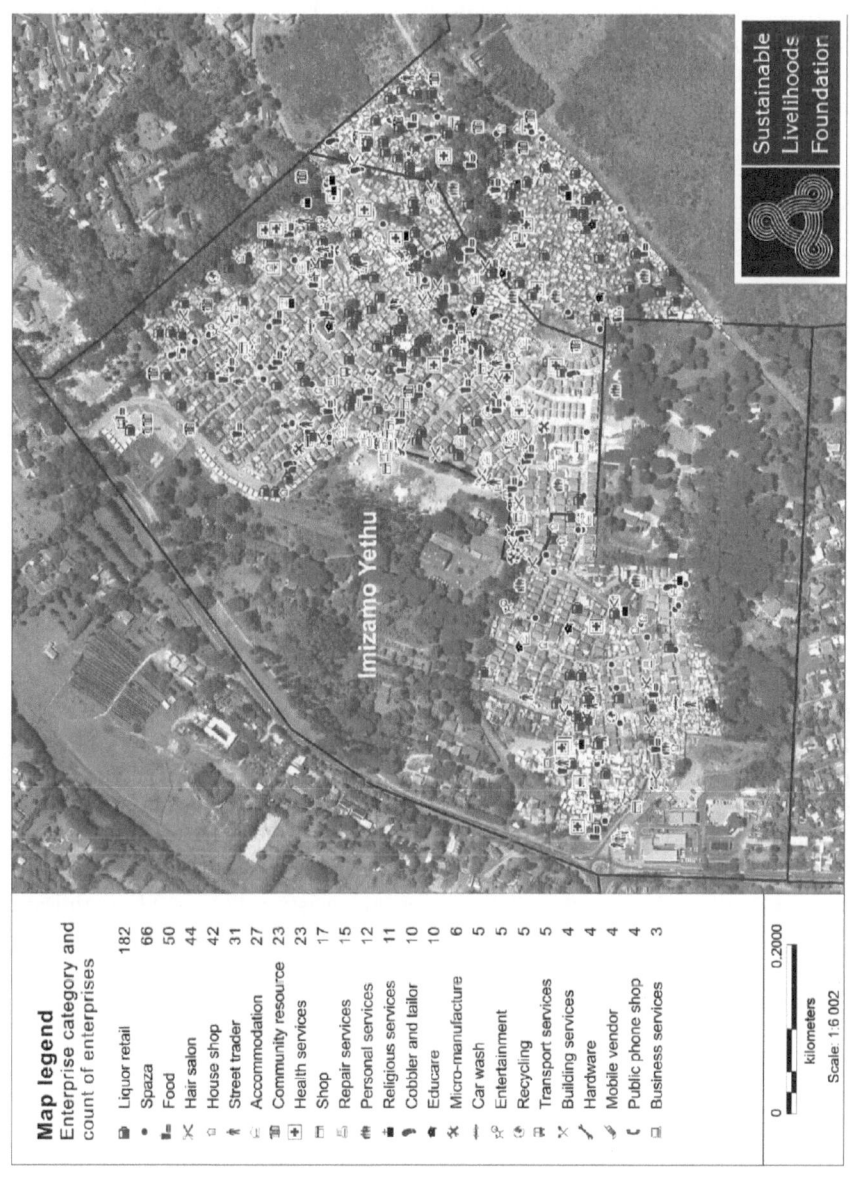

Map legend

Enterprise category and count of enterprises

🏠	Liquor retail	182
•	Spaza	66
▬	Food	50
✂	Hair salon	44
⌂	House shop	42
▲	Street trader	31
🏚	Accommodation	27
⊞	Community resource	23
✚	Health services	23
🏠	Shop	17
⚒	Repair services	15
⚘	Personal services	12
⛪	Religious services	11
🛠	Cobbler and tailor	10
✖	Educare	10
✗	Micro-manufacture	6
⚹	Car wash	5
✕	Entertainment	5
♻	Recycling	5
✗	Transport services	5
⟋	Building services	4
⟍	Hardware	4
⚲	Mobile vendor	4
☎	Public phone shop	4
▢	Business services	3

0   0.2000

kilometers

Scale 1:6 002

Imizamo Yethu

Sustainable
Livelihoods
Foundation

Figure 3.4 Distribution of informal businesses in Imizamo Yethu, January 2013 'world class city' attractive to investors, domestic and international, the City leaves no policy room for the urban poor who are 'a population that is superfluous to growth in the formal economy or embarrassing to those aspiring to world class city status'. This policy of 'forgetting' the urban poor results in what Charman, Tonkin, Denoon-Stevens, and Demeestére (2017: 1) identify as enduring forms of spatial development and land use management from the apartheid era that 'reinforce spatial injustice in the township rather than nurturing economic growth'.

This exclusion is manifested in City policy that leaves almost no place for poor residents to reside or pursue livelihoods legally. Thus, in 2012 the City of Cape Town adopted a Spatial Development Framework (SDF), wherein it has established guidelines about 'how and where Cape Town should grow in the future' (CCT 2012). The goals of the SDF are supported by the 'Cape Town Zoning Scheme' (CTZS) which appoints and formalises various land use designations on all landholdings. These include residential, agricultural, industrial, and commercial activities. The CTZS acknowledges differences between 'township' and 'suburb' residential areas, and 'in recognition of the realities of poor and marginalised communities, development rules ... are not very restrictive and local employment generation is encouraged' (CTZS 2012: 34). As per this accommodation, households in formally planned township residential areas are permitted to operate certain home-based business activities including house shops, childcare (up to six children), business services, religious businesses, and informal trading.

However, as Charman, Petersen, and Piper (forthcoming) argue, in reality the variegation in City policy is limited. Thus, the business activities allowed in the CTZS are additional to existing residential use and subject to certain constraints including floor space, signage, and limited opening hours. Buildings used for enterprise activities must have approved plans, a requirement that requires demonstration of land/property ownership and the use of approved building materials. Some business activities are prohibited explicitly, including video and gaming machines and liquor retailing. Furthermore, the CTZS excludes business activities that may present 'potential sources of nuisance', broadly defined, and may cause noise and waste pollution. The definition potentially extends to mechanical repairs, panel-beating, light manufacturing, glass and metal recycling, traditional beer brewing, live chicken sales, food preparation, appliance repairs and transport services. All such activities can be shut down at the City's discretion. Consequently, the space for emergent township businesses to formalise remains small and practically exclusionary, and the vast majority of informal enterprises currently evident in Imizamo Yethu exist in contravention of one or another of the City's by-laws.

A key point in this section is to illustrate how informal settlement in Hout Bay emerges from the intersection of, on the one hand, financially exclusionary market governance through the residential property market that dominates the Valley and, on the other hand, the limited supply of state housing through developmental governance in Imizamo Yethu and

Hangberg. Informal housing is thus a story of exclusion, and this extends to other forms of informality too, including livelihood practices linked to settlement. Despite various national laws and the City's policy initiatives around spatial planning, the tolerance of informal livelihoods remains low. In both its settlement and livelihood dimensions, exclusion has a profound effect on how poor people live in Hout Bay through its creation of grey spaces where formality and informality coexist. Thus, how people live in Imizamo Yethu is also an effect of the intersection of forms of governance with divergent logics in Hout Bay. In addition, informality has political expressions, too, that range from Holston's insurgent citizenship that challenges the state in the name of political equality and rights, to Bayat's quiet encroachment where residents avoid the state to pursue illegal activities until forced to defend themselves. More on these forms of politics in Chapters 6 and 7.

Hangberg: fractured by market, developmental, and informal governance

With Hangberg, we come full circle to the opening montage on Sentinel Mountain, and the tenure insecurity of poor people in the face of market governance of residential property. Historically the state has played the most central role in housing in Hangberg as in Imizamo Yethu. Thus, after the construction of the harbour in the 1950s, employers and the local government built residential flats (or hostels) for the 'coloured people' working in the local fishing industry. In the 1980s, brick row houses were built by local government to meet the need for increased labour (Rubin & Royston 2008). Today Hangberg has three distinct housing types: formal houses, most of which are owned by wealthy residents (mostly coloured but a few white) who live in the well-serviced area of Hangberg Heights; state-run hostels and rental housing inhabited by a poorer, mostly coloured community (which cost in the region of R200/per month for a two room house)[2]; and informal shacks (or bungalows as residents call them) at the foot of Sentinel Mountain, along the firebreak, or built in the backyards of rental housing units (Rubin & Royston 2008).

This threefold diversification of Hangberg is the best illustration of the cumulative and disordering impact of market and developmental governance on residential settlement in Hout Bay after apartheid. On the one hand, market governance has shaped the 'post-public' development of the Valley, as exemplified in the growth of gated communities; on the other, developmental governance has driven formal housing in Imizamo Yethu, the lack of which has seen the greying of residential property in the settlement. In Hangberg, however, the impact of both is evident. Thus, market governance is evident in the growth of the relatively opulent, middle-class area of The Heights, and developmental governance in the slow expansion of state-sanctioned housing in Hangberg itself. Lastly, since the turn of the century, informal housing has grown significantly, most famously on the firebreak above the settlement, leading to the Battle of Hangberg in 2010. This section will focus mostly on

Figure 3.5 Aerial image of Hangberg

the evolution of settlement patterns, and what it tells us about post-apartheid governance, with Chapter 4 exploring in more detail the politics manifested in the struggles around informal housing in Hangberg.

Since 1994, The Heights has grown in size but especially in value with the property market in Hout Bay rising more widely (see Figure 3.5). Located on the steep slope of a mountain that affords magnificent views over Hout Bay harbour and across to Chapman's Peak, it backs onto the protected Table Mountain Nature Reserve. Indeed, at some point in the mid-2000s it was apparently renamed the 'Harbour Heights', ostensibly to enhance its property values (Dogon 2007). Key to the attraction of the Harbour Heights is its location, and indeed Hangberg, too, is seen as 'prime land' that has potential value for many parties. Certainly, the settlement has commercial and land values that are not always apparent in other informal settlements (Rubin & Royston 2008). Indeed, when describing the new council residential units, a city official jokingly said he was 'reserving a seafront unit for himself as he would like to have a home with that view' and, given their location, they could potentially 'be worth a lot of money' (City official 2 2015). Western Cape Premier, Helen Zille, too described how the new council rental flats which are being built in Hangberg are on 'what is arguably the most beautiful piece of waterfront real estate in Africa' (Zille 2014). As one Hangberg resident explained,

> Land's got a massive value in South Africa, specifically in Cape Town. And ... if there's land close to the sea ... they call it prime land ... And maybe that's why they would ... always feel that poor people doesn't [sic]

need to stay next to the sea. They need ... to be [removed] somewhere or away from beauty. That is a perception that a lot of people got. So it's an economic aspect ... That's why I'm saying that the competitiveness for land, in specific on the coast lines, is quite ... controversial and quite intimidating towards the poor.

(Cited in Ehebrecht 2014: 128)

Between 1999 and 2007, the informal settlement on the upper slopes of Sentinel Mountain expanded under the auspices of the City Council, which provided each household with a letter of consent, thus granting de facto security of tenure (Fieuw 2011). Although the informal settlement received rudimentary services between 2001 and 2004, the residents remained generally unhappy with the level of services. In the early 2000s, for example, there were a total of 39 flush toilets and 37 water standpipes for a population of 400 households (Kapembe et al. 2007, in Fieuw 2011). As of 2008, the cost of a bungalow (or shack) ranged between R3,000 and R40,000 but prices increased to up to R50,000 once there was the promise of formal housing through the UISP programme discussed in the following chapter. In principle, what is being sold is not the land, but the shack and the number on the community register which promises a future house, development rights, and formal ownership.

Indeed the actual cost of housing in Hangberg is challenging to assess, as many residents do not own the land their homes are built on. Many of the informal bungalows are built on land owned by various bodies, including the municipal and Western Cape governments, and with claims by a private landowner to a part of the settlement. Nevertheless, the informal sale of properties is very much in evidence with bungalows being frequently bought and sold (Rubin & Royston 2008). In theory, these sales should be tracked on a community register that was set up under the auspices of the Hangberg in situ Development Association (HiDA), but this is no longer operational (Ehebrecht 2014). Given this scenario it is obviously not possible for buyers to access formal credit and leverage their investments (Rubin & Royston 2008). Thus, as in Imizamo Yethu, state housing in Hangberg follows the developmental logic of 'sheltering a population' rather than the market logic of conferring individual property rights. Further, and again similarly to Imizamo Yethu, residents of the informal sector in Hangberg are engaged in the marketisation of property through de facto renting, without the benefits of accessing traditional financing structures or the legal protection of property rights.

Positively, market forces are mediated by social norms, as Hangberg is a relatively small and tight-knit community. There were only 302 households in the informal settlement part of Hangberg entered into the original HiDA community register. Indeed the majority of those living in Hangberg come from families that have lived in the area for generations and thus households are often related in some way to each other (Tefre 2010). Consequently, residents only really access property through extended families and kin groups (Rubin & Royston 2008). Indeed, according to a leader of HiDA, the lack of capitulation to market forces is evident in that 'people are not greedy ... going after money by selling their houses the first opportunity they get' (DAG 2008: 19). In this sense, a house and land are seen as social goods rather than an individual commodity. Her response may be conditioned, however, by fear of existing Hangberg residents being exploited by 'the many corrupt people who realise the investment potential of the land' (DAG 2008: 19).

There is thus a significant impact on Hangberg of the marketisation of residential property in Hout Bay. The vast majority of Hangberg residents are not yet legal private property owners, as the pressure of the community and family expectations mediate the demands and pull of the market. Residents may start to succumb to 'downward raiding', however; that is, the temptation to sell property and land at inflated prices as soon as they have achieved secure tenure (Rubin & Royston 2008: 29). This may also happen as a result of capital subsidies that form part of the national housing strategy. Either way the demand for the scarce resources of land and housing forms an important context to the ongoing conflict in Hangberg. Insecure land tenure, fear of displacement, and enduring perceptions of 'insiders' versus 'outsiders' in state housing projects infuse the relations of place in Hangberg.

In short then, the 'greyness' of formality/informality in housing, evident in Imizamo Yethu, is reproduced in Hangberg too, with similar consequences: widespread 'backyarding' and infilling of public land, overloading of infrastructure, and increased risks to security, health, and the environment. This greyness is also evident in the job market where the recent regulation of the fishing industry has put many local fishers out of work, creating a thriving informal and criminal economy in fishing and perlemoen (abalone) smuggling. The production of livelihood informality and its consequences are discussed further in Chapter 9. For now, it suffices to note that informality applies in respect of livelihoods in Hangberg as well as housing, and is indicative of how the combined impact of market and developmental governance has made the poorer section of Hangberg a 'grey space' too.

Conclusion: the divergent meanings of housing in Hout Bay

The demise of apartheid and the advent of a non-racial democracy in 1994 promised the end of a legally entrenched racial system, including the segregation of the Group Areas Act. However, while the formal restrictions on who can access urban residential space have gone, they have been replaced by economic restrictions due to the distribution of most residential property through market means. The marketisation of city residential space is a trend throughout the world, even after the global crash of 2008, as property remains an excellent investment for financial capital. In addition, the DA government in Cape Town has been 'red carpeting' private development in the last 5 years.

This reality means that property developers set much of the agenda for city spatial development by defining where the middle classes live, shop, exercise, and socialise.

Further to defining where people live, market governance is also shaping how they live. Thus as Murray (2015) notes, private developers have pursued a particular imaginary of urban life, centred on security estates, redeveloped urban commercial zones, themed entertainment centres, and the new cathedral of urban consumerism: the mall. In the South African context this spatial and lifestyle power often overlays and reinforces, rather than challenges, the spatiality and exclusion of apartheid, such as the racial framing of crime. The apogee of neo-apartheid exclusion is the security estate, currently metastasising across the land mass of Hout Bay.

Given that land is a zero-sum public good, the rise of the residential property market has implications for poor residents too, with Harvey (2012) describing the segregation of urban spaces into wealthy enclaves and poor ghettos. Unable to control private developers and their shaping of the city or even backing them, post-apartheid local government has instead focused on servicing poor, black areas of the city through various forms of developmental governance. Consequently, where for wealthier groups in Hout Bay, housing is an issue of property through the market, for poorer groups, housing is shelter for the needy provided by the state. Thus market and developmental governance frame housing in divergent ways – on the one account it is about individual civil liberties under capitalist relations, on the other it is about collective needs met through the development state.

However, while South Africa has accomplished the largest delivery of housing in human history (Ferguson 2015), it is simply not enough to match the demand of poor migrants into the city. Squeezed out of (securitised) wealthy areas by poverty, and under-serviced by the developmental state, many poor residents have no choice but to build informal shelters, including in the backyards of existing state beneficiaries. Informality thus grows in poor areas, often within formal homesteads as well as around them, and before long many poor settlements come to resemble what Yiftachel (2009) terms 'grey areas' of both formality and informality, and of legality and illegality. Lastly, unlike green areas, grey ones are not well-organised, due to the limits of formal rule, but also because of the contradictory forms of governance that coexist within them. Thus, while in the Valley you can only get housing through the market, in Hangberg you can access it through market, state or informal means.

This constitution of the urban by divergent 'green' and 'grey' residential areas, archetypally represented in gated communities and informal settlements, with associated urbanisms as consumers and populations/ encroachers respectively, is what we term neo-apartheid. It is like apartheid in the way it segregates distinct social groups, hierarchically framed, into particular places in the city that are governed in different ways; indeed, there are strong continuities in the racial categories of inclusion and exclusion with the past, especially due to the economic legacies of apartheid. It is also like apartheid in that local residents have little democratic control over the allocation of various forms of resources, such as housing. However, there are important differences too, not just in the removal of legal racism, but also in the asymmetrical and slow racial desegregation of middle-class and wealthier areas though class mobility. A particular difference is the declining role of the state in meeting the needs and rights of the privileged, and its greater role in supplying the key needs of poorer communities. Thus, once again in South Africa, social hierarchies are spacialised, albeit in this case by money rather than race, and subjected to divergent forms of undemocratic governance.

Notes

1 The National Small Business Act of 1996 defines micro-enterprises as businesses employing <5 persons, that have a turnover of <R200,000 and assets worth <R100,000. Informal micro-enterprises are 'businesses that are not registered in any way … small in nature … and operated from homes, street pavements and other informal arrangements' (Statistics South Africa Labour Force Survey: March 2007).
2 Based on field notes, March–May 2015.

References

AfricaCheck. (2016). Democratic Alliance Wrong to Claim 67% of Cape Town's Budget Spent on Poor Communities. *AfricaCheck*, 14 June 2016, https://africacheck.org/reports/democratic-alliance-wrong-to-claim-67-of-cape-towns-budget-spends-on-poor-communities/. Accessed 30 September 2017.

Ashton, G. (2013). Alliance Between City and Developers Undermines Rights. *Letter to Cape Times*, 10 June. http://gctca.org.za/public-discourse-coct-attempts-to-undermine-public-participation/. Accessed 3 October 2016.

Ballard, R., & Jones, G. (2015). The Sugarcane Frontier: Governing the Production of Gated Space in KwaZulu-Natal. In Haferburg, C., & Huchzermeyer, M. (Eds.), *Urban Governance in Post-Apartheid Cities: Modes of Engagement in South Africa's Metropoles.* Pietermaritzburg: University of KwaZulu-Natal Press, 295–312.

BusinessTech. (2017). These are the Top 10 Richest Suburbs in South Africa. *BusinessTech*, 26 June. https://businesstech.co.za/news/wealth/181715/these-are-the-top-10-richest-suburbs-in-south-africa/. Accessed 30 September 2017.

Charman, A., & Piper, L. (2012). Xenophobia, Criminality or Violent Entrepreneurship? Violence against Somali Shopkeepers in Delft South, Cape Town, South Africa. *South African Review of Sociology*, *43*(3), 81–105.

Charman, A., Petersen, L., & Piper, L. (forthcoming). Informality Disallowed: State Restrictions on Informal Traders and Micro-Enterprises in Browns Farm, Cape Town, South Africa. Lagos: IPSS.

Charman, A., Tonkin, C., Denoon-Stevens, S., & Demeestére, R. (2017). Post-Apartheid Spatial Inequality: Obstacles of Land Use Management on Township Micro-Enterprise Formalisation. A Report by the Sustainable Livelihoods Foundation, 14 August. Cape Town: SLF. www.livelihoods.org.za/wp-content/uploads/2017/09/SLF_Post-Apartheid_Spatial_Inequality_-_lr_version.pdf. Accessed September 2017.

CCT (City of Cape Town). (2012). *Cape Town Spatial Development Framework Technical Report*. http://resource.capetown.gov.za/documentcentre/Documents/ City%20research%20reports%20and%20review/SDF_Technical_Report_2012_ Interactive.pdf. Accessed 1 August 2017.

CCT. (2015). *City of Cape Town Municipal Planning By-Law*. www.westerncape.gov. za/eadp/files/basic-page/uploads/City%20of%20Cape%20Town.pdf. Accessed 30 September 2017.

CCT. (2016). *Five-Year Integrated Development Plan 2012–2017. 2016/17 Review and Amendments*. www.capetown.gov.za/Family%20and%20home/meet-the-city/our-vision-for-the-city/cape-towns-integrated-development-plan. Accessed 3 October 2016.

CCT. (2017). *City of Cape Town 2017/18–2019/20 Budget*. Media briefing by Councillor Johan van der Merwe, Mayoral Committee Member for Finance. 29 March 2017. http://resource.capetown.gov.za/documentcentre/Documents/Graphics%20 and%20educational%20material/Draft%20Budget%202017_18_Media%20 briefing%2029March2017_FINAL.pdf. Accessed 30 September 2017.

CTZS (Cape Town Zoning Scheme). (2012). *City of Cape Town Zoning Scheme Regulations: A Component of the Policy-Driven Land Use Management System*. CCT: Cape Town. http://sans10400.co.za/wp-content/uploads/2013/06/Cape-Town-Zone-Scheme_Regulations_Nov_2012_Part1.pdf. Accessed 30 September 2017.

DAG (Development Action Group). (2008). *Lessons Learned in Leadership: Case Studies from Development Action Group's 2008 Community Leadership Programme*. www.dag.org.za/images/pdf/case-studies/2008_Case%20Study_Lessons%20 in%20Leadership%20from%20Community%20Leadership%20Programme.pdf. Accessed 19 November 2015.

De Soto, H. (2002). *The Other Path: The Economic Answer to Terrorism*. New York: Basic Books.

Dogon. (2007). Recent Sales Show Strong Demand for Hout Bay Homes. *Dogon Group Properties*, 13 June. www.dogongroup.com/press/481-recent-sales-show-strong-demand-for-hout-bay-homes.html. Accessed 3 October 2016.

Ehebrecht, D. (2014). The challenge of informal settlement upgrading: breaking new ground in Hangberg, Cape Town? Master's thesis, Universitätsverlag Potsdam, Potsdam.

Ferguson, J. (2015). *Give a Man a Fish. Reflections on the New Politics of Distribution*. Durham, NC: Duke University Press.

Fieuw, W. V. P. (2011). Informal settlement upgrading in Cape Town's Hangberg: local government, urban governance and the 'right to the city'. Master's thesis, Department of Philosophy, Stellenbosch University.

HAA. (2001). *Housing Amendment Act, No 4 of 2001*. www.saflii.org/za/legis/num_ act/haa2001187.pdf. Accessed 3 September 2017.

Harte, W., Sowman, M., Hastings, P., & Childs, I. (2015). Barriers to Risk Reduction: Dontse Yakhe, South Africa. *Disaster Prevention and Management*, *24*(5), 651–669.

Harvey, D. (2012). *Rebel Cities: From the Right to the City to the Urban Revolution*. London & New York: Verso.

Huchzermeyer, M. (2011). *Cities with Slums: From Informal Settlement Eradication to a Right to the City in Africa*. Cape Town: UCT Press.

Jackson, K. (1985). *Crabgrass Frontier: The Suburbanisation of the United States*. Oxford: Oxford University Press.

Johnson, T. (2016). Cape Town's Development Frenzy. *Politicsweb*, 14 July 2016. www.politicsweb.co.za/opinion/cape-towns-development-frenzy. Accessed 17 October 2016.

Lefebvre, H. (1991). *The Production of Space*. Oxford: Blackwell.

Lemanski, C. (2004). A New Apartheid? The Spatial Implications of Fear of Crime in Cape Town, South Africa. *Environment & Urbanisation, 16*(2), 101–112.

Lemanski, C., Landman, K., & Durington, M. (2008). Divergent and Similar Experiences of 'Gating' in South Africa: Johannesburg, Durban and Cape Town. *Urban Forum, 19*(2), 133–158.

Moneyweb. (2010). Hout Bay's Sentinel Mountain Still for Sale. But Where are the Buyers? *Moneyweb*, 25 May. www.moneyweb.co.za/archive/hout-bays-sentinel-mountain-still-for-sale/. Accessed 3 October 2016.

Murray, M. (2015). City of Layers: The Making and Shaping of Affluent Johannesburg after Apartheid. In Haferburg, C., & Huchzermeyer, M. (Eds.), *Urban Governance in Post-apartheid Cities: Modes of Engagement in South Africa's Metropoles*. Pietermaritzburg: University of KwaZulu-Natal Press, 179–198.

News24. (2009). Bid to block mountain sale. *News24*, 5 August. www.news24.com/SouthAfrica/News/Bid-to-block-mountain-sale-20090805. Accessed 5 October 2016.

Property24. (2016). Hout Bay Trends and Statistics. *Property24*, October. www.property24.com/hout-bay/property-trends/615. Accessed 5 October 2016.

Propertywheel. (2016). Hout Bay Property Market Sees a Boom. *Propertywheel*, 01 February. http://propertywheel.co.za/2016/02/hout-bay-property-market-sees-a-boom/. Accessed 5 October 2016.

Rubin, M., & Royston, L. (2008). Scoping Study – Local Land Registration Practices in South Africa. *Urban Landmark*. www.urbanlandmark.org.za/downloads/Local_Land_Registration_Practices.pdf. Accessed 30 September 2017.

SERI (Socio-Economic Rights Institute of South Africa). (2013). 'Jumping the Queue': Waiting Lists and Other Myths. Perceptions and Practice Around Housing Demand Allocation in South Africa. Community Law Centre, UWC & SERI, April. www.seri-sa.org/index.php/more-news/174-research-report-jumping-the-queue-waiting-lists-and-other-myths-perceptions-and-practice-around-housing-demand-and-allocation-in-south-africa-3-july-2013. Accessed 30 September 2017.

Swimmer, L. (2013a). Another potential blow to democratic local government. Letter to Mayor de Lille 11 June 2013. *Greater Cape Town Civic Alliance*. http://gctca.org.za/public-discourse-coct-attempts-to-undermine-public-participation/. Accessed 3 October 2016.

Swimmer, L. (2013b). HBR&RA's Land Use/Planning Portfolio Report. *Hout and About, HBRRA Newsletter*, July 2013. www.houtbay.org.za/RAHB_Newsletters.html Accessed 3 October 2016.

Swinhoe, D. (2012). Cape Town: A City with a Vision. *The South African*, 17 April 2012. www.thesouthafrican.com/cape-town-a-city-with-a-vision/. Accessed 3 October 2016.

Tefre, Ø. S. (2010). Persistent inequalities in providing security for people in South Africa – a comparative study of the capacity of three communities in Hout Bay to influence policing. Master's thesis, Department of Administration and Organization Theory, University of Bergen.

Winde, A. (2011). The W. Cape govt's plan to put red tape – Alan Winde. *Politicsweb*, 4 August 2011. www.politicsweb.co.za/documents/the-wcape-govts-plan-to-cut-red-tape--alan-winde. Accessed 3 October 2016.

Yiftachel, O. (2009). Critical Theory and 'Gray Space': Mobilization of the Colonized. *City*, *13*(2–3), 246–263.

Zille, H. (2014). The Salutary Story of Hangberg. *The Gremlin*. www.thegremlin.co.za/2014/03/24/the-salutary-story-of-hangberg/. Accessed 30 September 2017.

Interviews

1. Air pollution activist (2017). Interviewed by Fiona Anciano. 17 August 2017.
2. City official 1. (2015). Interviewed by Laurence Piper. 7 April 2015.
3. City official 2. (2015). Interviewed by Fiona Anciano. 30 March 2015.
4. Community leader 2. (2012). Interviewed by Laurence Piper. 10 May 2012.
5. Community leader 2. (2017). Interviewed by Laurence Piper. 24 August 2017.
6. Denoon-Stevens, S. (2017). Former official in planning Department, City of Cape Town, 2010–2012. Interviewed by Laurence Piper and Fiona Anciano. 22 August 2017.
7. Environmental activist 1. (2017). Interviewed by Laurence Piper. 24 August 2017.
8. Estate agent 1. (2015). Interviewed by Fiona Anciano. 15 April 2015.
9. Estate agent 1. (2017). Interviewed by Laurence Piper. 22 August 2017.
10. Estate agent 2. (2018). Interviewed by Laurence Piper. 24 January 2018.
11. PMF. (2017). Focus group with first leaders of Hangberg Peace and Mediation Forum (PMF). Interviewed by Fiona Anciano and Laurence Piper. 24 August 2017.
12. Provincial officials. (2015). Interviewed by Laurence Piper and Fiona Anciano. 27 March 2015.
13. Trader. (2017). Interviewed by Laurence Piper. 25 August 2017.

4 Defending the shack

The politics of developmental governance

On a quiet day in September 2010, in the small mountainside community of Hangberg, the City of Cape Town's anti-land invasion unit arrived to demolish shacks that had been erected on Sentinel Mountain. An eviction order had been issued to dismantle nominally 'unoccupied' shacks as they were illegally built on a firebreak. Residents of Hangberg, however, described how many of the shacks *were* occupied, and filled with furniture and personal items. A group of residents gathered to block the land invasion unit, and in response, the City then called in the police. Within hours, the Hangberg area had become a battleground, with dozens of residents resisting police invasion, often while young children looked on. The running battles lasted for several hours with police firing rubber bullets and residents retaliating by throwing rocks and homemade petrol bombs (see Figures 4.1 and 4.2). There were serious injuries on both sides; three Hangberg residents each lost an eye due to police bullets. Beyond the physical injuries, this was a community left traumatised, shocked that their city would deploy police against them when in their view they were simply protecting their homes (Kaganof & Valley 2010).

The Battle of Hangberg represents a defining moment in the history of Hout Bay. It is an event, however, that is steeped in a long history of dialogue and confrontation with the state. This chapter unpacks the events that led to the conflict and the forms of governance resulting from it. To restore peace to Hangberg, the courts mandated the formation of a Peace and Mediation Forum (PMF) to be the exclusive structure to represent Hangberg to the state in all future development projects in the community. This forum, understood as a form of developmental co-governance, has contributed much to maintaining the peace, but little to meaningful development.

Framed as a partner with the state in co-governing development in Hangberg, the PMF has no real powers, no resources, and poorly defined electoral and organisational processes. Consequently, it is very much a junior partner, overwhelmingly reliant on the state to define and drive development in Hangberg. This has allowed the state to choose its commitments and to implement them at its own pace. Conversely, the weak organisation of the PMF has seen changes in leadership without clear processes, undermining its efficacy and legitimacy. At the same time, other organisations are prevented

Figures 4.1 Images of the Battle of Hangberg (Photos by Stephen Williams)

Figure 4.2 Images of the Battle of Hangberg (Photos by Stephen Williams)

by law from engaging the state on development projects, profoundly limiting the role of civil society in the development of Hangberg. In this context, developmental governance has meant state *dirigisme*, the (alleged) capture of the PMF by a local network of individuals, and the replacement of politics with new forms of patronage and nepotism.

The story of Hangberg

Situated above the Hout Bay Harbour, Hangberg was zoned as a coloured area under the 1950 Group Areas Act. Coloured inhabitants of the Valley, including those of Khoi/San descent, were forced to move to Hangberg by 1956. The area allocated for Hangberg residents is a small percentage of the larger Hout Bay Valley and, although it has beautiful views of the bay, it is situated on a steep side of the Karbonkelberg Mountain (Fieuw 2011; de Greef 2013). Two chief issues have brought state and society into conflict, and at times co-operation, in Hangberg. The first is ongoing and systematic poaching of rock lobster and abalone by a well-organised fishing community, which will be discussed further in the following chapter; and the second, the focus of this chapter, is land ownership and housing.

As discussed in Chapter 3, Hangberg's housing has developed over many decades through market, state, and informal means resulting in a range of housing types. These are formal, large houses owned by wealthy residents (both coloured and white) in the well-serviced area of Hangberg (Harbour) Heights; state-owned hostels and rental housing in Hangberg proper, inhabited by poorer residents; and lastly, informal shacks (or 'bungalows') built by the poorest of the poor along the firebreak at the top of Hangberg, or in the back-yard of rental housing units (Rubin & Royston 2008; see Figure 4.3).

As outlined in the preceding chapter, a key factor that led to the Battle of Hangberg was ongoing insecurity over access to housing, land, and land

Figure 4.3 Image of Hangberg and the Heights with housing type

tenure in Hangberg. Given its breathtaking views of the Hout Bay Harbour and world-famous Sentinel Mountain, Hangberg occupies a premium position geographically. Hangberg is seen as 'prime land' with potential value not always apparent in other poor or informal settlements (Rubin & Royston 2008: 29). As noted in the preceding chapter, the actual cost of housing is difficult to assess given the great variety of forms it takes. One of the outcomes of the relatively high premium on land in Hangberg, and the lack of secure tenure, is a pervasive fear of displacement, manifested in the protest against the 'sale of the mountain' in 2009, described in Chapter 3.

This fear is not baseless. In the early 2000s there were tentative plans to relocate the community to an infamous area on the outskirts of Cape Town called 'Blikkiesdorp' (Tin Can Town) in Delft South, due to the difficulty of upgrading the area (Bakkes 2011; Ehebrecht 2014). These concerns surfaced after comments from the newly-elected ward councillor, Marga Heywood, in 2007, who stated that 'Hout Bay is in a big mess and the only solution I see is that people are removed from here – even if it is against their will' (Joubert 2007: n.p.). A community leader from Hangberg, Timothy Jacobs (quoted in Joubert 2007: n.p.), explained that this was not a viable option:

> Our people will not move. Our parents were forcibly removed from what is today the white Hout Bay. Now people in shacks have to move again. It will not happen. Black and coloured people have access to only 4% of land in Hout Bay. The city must find a solution here in Hout Bay to deal with the housing shortages. There'll be war before we move.

His words were prescient. The informal settlement at the top of Hangberg began shortly after 1994, and is populated mostly by people with family connections in the settlement. Initially the City Council provided each household with a letter of consent, thus granting de facto security of tenure, and committed to providing some services. However, this did not stop the further spread of shacks, and while some rudimentary services were provided by the City between 2001 and 2004, they were insufficient. As the informal settlement grew without enough clean water, sanitation and electricity, the community engaged the City about service provision (Fieuw 2011). In March 2007, the City of Cape Town successfully applied for the country's first in situ upgrade via the Upgrading of Informal Settlements Programme (UISP).

Initially there was optimism that the UISP would be successful as it suits environments where there are high levels of community-initiated development and relatively contained, tight-knit communities with strong leadership and civil society structures that can navigate the complex negotiations required. These dynamics were, it appeared, present in the case of Hangberg (Fieuw 2011). The Hout Bay Civic Association (HBCA) – a Hangberg-based civil society organisation historically linked to the ANC – worked closely with an NGO called the Development Action Group (DAG) to lobby for the UISP. There was also hope that the UISP would put paid to concerns of eviction as the Programme included the principle that there would be a prevention of market-related displacements and no relocations out of Hangberg. In theory then, UISP promised to regulate land ownership and grant residents security of tenure (Ehebrecht 2014).

From the point of view of the state, UISP also looked promising, as it would include those who do not qualify for housing subsidies, ensuring they would not be displaced, and protect existing social networks. It would allow for a more flexible approach to tenure solutions and variation from planning norms to allow for in situ upgrading. In particular, the City intended to use UISP to provide sanitation, potable water and access paths as well as subdividing the area, providing security of tenure to informal settlement residents. The City also hoped to ensure that the residents of the informal settlement were registered and to prevent further settlement growth. In short, the City planned 'to use municipal funds to improve informal settlements in partnership with their residents. And to do that incrementally step by step. That's our strategy' (City housing consultant cited in Ehebrecht 2014: 98).

To support the UISP, a project steering committee was formed in Hangberg, called the Hangberg in situ Development Association (HiDA). This consisted of four steering members and representatives from every block in the informal settlement. They were elected democratically to represent the community in matters pertaining to the upgrading project (Development practitioner 2017). By October 2007, HiDA and the City of Cape Town had agreed on a moratorium on the construction of new informal structures in Hangberg. If the moratorium were broken, HiDA would ask City law enforcement to intervene. The agreement also discussed the scope of the UISP, limiting the project to 302 housing opportunities, as per the number of informal households on a register drawn up by HiDA.

The City, however, did not set up any arrangement to monitor the moratorium and so HiDA was left with the difficult task of policing the agreement themselves through negotiations with local residents. Furthermore, new power dynamics materialised between the qualifiers and non-qualifiers for the proposed development, creating new opponents to the upgrading (Ehebrecht 2014; Fieuw 2011). This new dynamic was exacerbated by the perception that the leaders of HiDA were misappropriating funds. As a KhoiSan Chief (2015) and resident of Hangberg explained, 'a lot of money never came forward and it disappeared'. A long-standing leader of a civil society organisation in Hangberg, described how 'leaders had access to opportunities ... they sold out the community for themselves' (PMF leader A 2015).

The Battle of Hangberg

Three years after the signing of the moratorium on the construction of new informal structures in Hangberg, housing concerns erupted again. The attempted auction of Sentinel Mountain, described in Chapter 3, created further land insecurity among informal settlement residents in particular.

Figure 4.4 Shacks being dismantled in Hangberg (Photo by Stephen Williams)

There was anger that they were not allowed to move above the firebreak, when the very land could (ostensibly) be sold to a developer. While this was not actually the case, and the land was sold to South African National Parks (SANParks), this misconception helped fuel anger among Hangberg residents. Thus midway through 2009, partly in protest against this perceived double standard, families started moving and building shacks above the fire-break, breaking a key agreement in the moratorium between HiDA and the City (Development practitioner 2017).

The previous Mayor of Cape Town and then Premier of the Western Cape, Helen Zille, attempted to address the infringements of the 2007 agreement by holding a public meeting that was ultimately unsuccessful. Following this, Zille ordered the city police to dismantle all 'unoccupied' structures on the firebreak (see Figure 4.4). Claiming the shacks were unoccupied gave more legitimacy to the police action. Members of HiDA were very frustrated as they felt they had not been given time to find a solution before the police arrived. Residents were clear that these structures were in fact occupied, and thus on 21 September 2010 there was a violent confrontation between police, representing the City, and residents, mainly from the informal settlement, in what came to be known as the 'Battle of Hangberg'.

The battle between police and residents waged for two days with numerous arrests and several significant injuries to residents caused by rubber bullets, including the loss of sight for at least three people (Kaganof & Valley 2010). While there are different views on what triggered the violence, it was the use of force by the police that resulted in these serious injuries. For Western Cape Premier Helen Zille, the violence came 'from a small group primarily known as the Rastas. They are both drug users and often, I'm afraid, drug peddlers' (Majavu 2010). Interviews and evidence from residents of Hangberg suggest that the main reason for conflict was rather that residents on the firebreak tried to prevent their shacks from being demolished (Kaganof & Valley 2010; Buhler 2014).

Notwithstanding the extensive media coverage of the 'Battle of Hangberg', the City of Cape Town was not deterred, and later in the month applied for a court eviction order to dismantle a further 52 shacks on the firebreak. However, SANParks, who own the land above the firebreak, and who had previously endorsed the clearing of shacks, withdrew support for the eviction order. This change of heart appeared partly politically motivated, an attempt by the ANC, which runs the national government including SANParks, to tarnish the reputation of the DA (SAPA 2010). The court, with the backing of relevant parties, decided not to issue an eviction order, instead mandating the formation of a mediated forum to discuss the way forward in the upgrading project. Through this court order, the Peace and Mediation Forum (PMF) was initiated.

The formation of the Peace and Mediation Forum

A mediator was appointed by the City of Cape Town to facilitate the court order. He explains how 'the first stage of the peace and mediation process was to get a single integrated community leadership established' which was made possible via a community meeting where a mandate was obtained 'to elect grass roots geographically situated sector based leaders. In this way, leaders from different parts of the community could be elected to represent the different interests and needs' (Mediator 1 2017). Over a period of seven weeks, residents from three sectors in Hangberg (informal residents, rental dwellers and ratepayers) elected 39 leaders. A core steering committee and representatives from every block were chosen. Most Hangberg residents supported the elections, but there was some contestation. A leader of the PMF explained how some leaders of the HBCA were elected onto the PMF, while others resented the formation of the PMF and contested its legitimacy, distancing themselves from the PMF process (PMF leader A 2015).

Once representatives were elected, it was possible for the PMF to represent the Hangberg community in signing a Peace and Mediation Accord with SANParks, the Western Cape Provincial Government, and the City of Cape Town. The Accord became an official order of the court in September 2011. It focused on the need for those living above the firebreak to move down into the settlement and outlined measures to support this. It also stated that the PMF could be called upon to assist with compliance to this measure, similar to the role HiDA had previously occupied (HPMA 2011). According to then Cape Town Mayor, Patricia De Lille (2011):

Those who move beneath the firebreak will be encouraged to move into areas that we can provide services to. ... Still others will move into new housing opportunities once developed ... The whole community ... will especially receive the benefits of the City's departments of Economic Development, Social Development and Sports Departments as well as the Department of Human Settlements, to say nothing of general City services.

The Accord included a range of state-provided programmes that the City, SANParks and the Province were mandated to work on with the PMF. The City in particular was legally obliged to implement programmes from a wide range of departments, including Economic Development, Environmental Health, Social Development, Human Settlements (including the UISP programme), and Sports and Recreation. These included broad-ranging aims such as 'to improve the capacity of the communities to champion and sustain economic development initiatives', as well as programmes to address 'Early Childhood Development, Youth ... Targeted Poverty Alleviation, Substance abuse, Violence and gangsterism, organised criminal activity (poaching or marine resources), the prevalence of anti-social subcultures' and 'domestic violence'. SANParks, too, was mandated to provide services such as improved opportunities for tourism, and the maintenance of cultural heritage. The Western Cape Provincial government had to run programmes on economic development, education (including literacy and numeracy interventions in the schools), social development, and health and community safety (HPMA 2011).

The Hangberg Peace Accord was clearly wide-ranging in both the scope of its programmes, and in including multiple spheres and organs of government in economic, social and infrastructure projects. As one PMF leader explained, a lot of work and deliberation went into framing the Accord, with potentially important outcomes that would deal 'not only with land and housing but also social and educational development for Hangberg' (PMF leader A 2015). For the 39 members of the PMF who signed the document it may have seemed like they were about to get comprehensive and wide-ranging services from government that could truly transform their community. However, as of 2018 this transformation has yet to occur.

The PMF and developmental governance

The formation of the PMF triggers interesting questions about the nature of urban governance, specifically in environments of contestation and scarcity, as exists almost everywhere in the urban South. While the City's role in implementing the orders of the Accord is a form of developmental governance, in conjunction with other spheres of the state and the PMF, it constitutes a form of co-governance. As the PMF mediator explains, 'The Peace Agreement is unique in that its non-compliance can result in criminal sanction. The High Court is an enforcement agency' (Mediator 1 2017). The state, particularly City Hall, is legally compelled to work with the PMF, and only the PMF, to deliver services and projects to Hangberg. The PMF can and must set the agenda of local government service delivery in Hangberg. Indeed the Accord states that:

> The Parties commit to ongoing engagement and liaison ... and to jointly seek solutions ... in particular the City and the Province shall meet with the Peace and Mediation Forum leaders at least every 2 months or as agreed.
>
> (HPMA 2011: Article 9)

Hence, the Peace and Mediation Accord is an example of an attempt at developmental co-governance, both in the sense of bringing development to Hangberg but also in the model of partnership between spheres of the state and community representatives in revising and implementing development plans. In a sense, the PMF has 'power with' City Hall to set the agenda for governance, but City Hall retains 'power to' deliver required services and projects (see Stone 1989, 2006). In fact, in this relationship only the state can produce the outcomes of the developmental partnership. Ackerman (2004) describes effective co-governance as a practice that essentially embeds societal actors in the process of governing from the outset of a development process. This potentially generates heightened forms of accountability that go beyond holding elected officials to account and hold everyday bureaucrats accountable too.

According to Ackerman (2004: 451), co-governance builds on ideas relating to co-production but deepens the accountability and participatory elements of co-production. Thus, 'in addition to coproducing specific services and pressuring government from the outside, societal actors can also participate directly in the core functions of government itself'. Ackerman places a positive democratic framing on ideas of co-governance, arguing that if carefully applied, co-governance can be much more rewarding than alternatives such as marketisation, and bureaucratic insulation:

> 'co-production' or 'societal accountability' ... by transgressing the boundaries between state and society institutional reformers can unleash invaluable pro-accountability processes which are almost impossible to tap into through less ambitious strategies.
>
> (Ackerman 2004: 460)

Did the model of developmental co-governance work?

How successful then has this form of co-governance been? Has the PMF been able to generate the breadth and depth of socio-economic development set out in the Accord and have multidimensional arms of the state worked effectively with the PMF? From the perspective of engagement and communication, there certainly has been success. Both the City and PMF members agree that there has been constant communication between the City, Province, and the Forum. The Premier of the Western Cape explains (Zille 2014),

> There is no community in Cape Town – not a single one – that has received the ongoing attention of the Mayor and myself, or the sheer commitment of resources, that Hangberg has. We have met with elected representatives of the community at least 30 times in scheduled and unscheduled engagements (and at least 300 days have been consumed by ongoing mediation and facilitation) … and we have a dedicated task team of personnel from the City and the Province to detect and monitor the progress of implementing the Peace and Mediation Accord.

Correspondence from the Western Cape Provincial government (2017) confirms this, explaining that:

> Community engagement with official Hangberg structures has been essential in how the City and Province have addressed deliverables of the Accord. The working committee serves as a great platform to coord-inate the relationship, progress, feedback and concerns related to projects and deliverables of the Peace Accord. … The City and Province work extremely closely with the HPMF.

Beyond communication, however, actual project delivery has been slow and limited to forms of housing. Thus, in terms of the UISP programme and the upgrading and provision of housing, City-owned rental units have been upgraded (according to one PMF leader at a cost of one million Rand each) and a new block of flats was built in 2016 (Figure 4.5). Beyond this, however, developmental co-governance has delivered little to nothing.

Even in respect of the building of one set of flats, neither the PMF, City, nor Province would claim that the Peace and Mediation Accord has been suf-ficiently implemented. The flats were promised in 2006, yet only manifested in 2016. Ten years for a small-scale development project involving 302 households is hardly a success story. The second block of flats promised in the Accord is still to be built, largely due to concerns with air pollution emanating from the nearby fish factory, which led the proposed development to fail an Environmental Impact Assessment (Air pollution activist 2017). This failure is perhaps best captured in Western Cape Premier Zille's (2014) comments that:

> [I]t was far easier to deliver a World Cup stadium, the biggest-ever infra-structure project in Cape Town, than it is to even get started on a modest housing project. Cape Town Stadium was completed within two-and-a-half-years. … But housing projects initiated as early as 2006, are still limping along.

Other aspects of the Accord, particularly those dealing with economic oppor-tunities and social issues, have been even less successful. As Chapters 5 and

Figure 4.5 New community residential units

9 explain, there are ongoing social and economic concerns in Hangberg. Poaching, or protest fishing, is prevalent and brings with it links to international criminal syndicates and drugs. It also leads to social problems in the community. There are relatively high levels of domestic abuse, poor schooling records, and ongoing social tensions (Pastor 1 2015; Political representative 1 2017; SSP patroller 2016). Indeed, some PMF leaders feel the state and the City have let them down, in particular by not fulfilling the social development aspects of the 2011 Accord. A PMF leader explained that there has been no significant social development programme in Hangberg, which is in desperate need of assistance with drug and alcohol addiction, specifically among the youth (PMF leader A 2015). They also feel bureaucrats in particular do not work to the Accord in Hangberg, but rather to the wider policy imperatives of the City, leaving important elements of the Accord unfulfilled (PMF leader B 2015).

The challenges confronting developmental co-governance in Hangberg

From the perspective of the state, the limited gains in achieving the commitments of the Peace and Mediation Accord stem in significant part from division and rivalries in the Hangberg community that undermine project implementation. Thus Premier Helen Zille believes that, 'had the community worked together with the City and the Province and stuck to negotiated agreements' progress would have been faster. Zille (2014) attributes problems in Hangberg to the idea that for the community it is,

> easier to remain marginalised and reinforce the perception of victimhood, than it is to seize the extraordinary opportunities and take responsibility for the development and transformation of their community ... The Mayor and I have often been tempted to withdraw from the Hangberg process altogether, and concentrate on communities more amenable to co-operative development.

There is clearly something to this analysis, especially as regards the divisions within the Hangberg community that the PMF failed to resolve. At the same time, it is important not to confuse effect for cause, as there is a strong case that the leadership conflicts and political divisions are a product of the politics of the PMF and developmental governance as much as a cause. There are three key elements to this: partisan contestation of the PMF's role; the poor design of the PMF as an institution of co-governance; and the resultant dependency on state delivery and resultant forms of patronage politics. Unable either to hold the state accountable or to consolidate its legitimacy with residents, the PMF has become a rubber-stamp for state projects. At the same time, its very existence prevents other civil society formations initiating development projects with the state, thus closing down space for innovation beyond the state.

Partisan contests of the PMF role

As already noted, the legitimacy of the non-partisan PMF was contested from the outset by the ANC-aligned HBCA. Thus, at a meeting in 2011 called by the PMF to explain the details of the Accord to affected residents, PMF leaders were heckled by the HBCA and unable to talk to attendees. The HBCA were offended that not everyone seemed equally free to attend the meeting and that attendees had to write down their names and sign in (Bakkes 2011). According to an HBCA member, 'the PMF was just an elite elected group and they never speak to people on the ground' (HBCA member 1 2015). This view is contested by several of the original elected leaders of the PMF who stated that those aligned to the ANC did not 'get chosen by the community' because 'they always had a political agenda, nothing to do with the community'. They tried to disrupt the PMF because they had lost their ability to control resources (PMF 2017).

The HBCA also took a different approach to shack evictions by the City than the PMF. The PMF, like HiDA, are in a difficult position as the Accord does not allow for further building on the firebreak, but the HBCA strongly feel that those who build houses on the firebreak should not be evicted. After the arrest of a resident in 2014 who had built a structure in 2010 (allegedly before the Accord), for 'breaching a high court order', there were protests that damaged cars in front of a block of flats where one of the PMF leaders lived. Anger was directed towards the PMF because, according to an HBCA member, 'the PMF knew that cops is gonna come in here, they never told the people, they are actually creating more problems' (HBCA member 1 2015). Roscoe Jacobs, leader of HBCA, said arrests of community members were a:

> huge miscarriage of justice. … People have become very frustrated. The arrest on Tuesday morning is the fourth or fifth since last year. The court order stems from 2012 for people who built their structures illegally in 2009 and 2010 … How long will this court order from 2012 be valid for? There is a lot of uncertainty in the community and the only reason they continue to build is because the city and provincial government did not address their issues properly.
>
> (Bezuidenhout & Williams 2014: n.p.)

Tensions in the composition and functioning of the PMF eventually went beyond the HBCA.

Indeed, in 2013 the PMF split when members disagreed over several issues, including negotiations with the City and differences of view with the Mayor and Premier. The Forum was, according to a local Pastor in Hangberg, reconstituted through new elections at the end of 2013, although it now has a smaller leadership component (Pastor 1 2015). According to an HBCA member, they re-elected only ten of the 30 leaders (HBCA member 1 2015) while the PMF says there are 'about 20' active committee members (PMF leader B 2015). To understand this better, we need to unpack the poor design of the PMF as an institution of co-governance.

Poor institutional design

The poor design of the PMF applies both in its capacities to co-govern development projects and to legitimate itself with residents of Hangberg. As noted, the PMF is empowered by the Accord to engage the City and Province regularly around programme implementation, and the failure to implement potentially invokes sanction by the High Court (HPMA 2011). However, within this process there is a profound asymmetry between civil society leaders and state officials in their capacity to act on these powers. Thus, unlike officials, the leaders of the PMF are not paid for their time and efforts; they must co-produce development in their free time after work. With one or two exceptions, PMF leaders are not well-qualified professionals, unlike most state officials, and have limited exposure to the technical knowledge and procedures of City and provincial line departments. Finally, the power of legal accountability exists in principle but requires expensive and expert legal capacity to be exercised in practice. It is thus unrealistic to expect working people volunteering their time to engage as equals with state officials in the terrain of developmental governance. Indeed, it is hard to see how the PMF could work as a space of co-production of developmental projects in its current design. At best, it can be a consultative forum dominated by state officials. In addition, only the state can actually implement the decisions taken as a result of this process.

At the same time as the PMF is authorised to speak for all residents of Hangberg, it has a weakly developed organisational structure, with no thoroughly specified rules around leadership authorisation and succession. There is also no set of rules that outlines standard operating procedures, organisation functioning, and the like. In short, as an organisation, leaders can legitimate themselves with residents of Hangberg, and work in an effective manner through a few well-defined and widely recognised procedures. This weak organisational framing means that leadership positions are easy to contest by invoking some form of democratic election, and in fact are not that different from other leadership positions in local communities that must be continually defended against rivals.

To return, then, to the leadership split in the PMF in 2013, a key reason behind this was that six of the core women originally elected onto the PMF when the Accord was signed were forced off the Forum. The women explained that they had started to challenge the City and Province for not providing the promised social support and economic development, particularly around issues such as domestic violence and alcoholism. After ongoing frustration with the state's lack of action, they wrote an email to the City and Province saying that they would leave the Forum if the City did not uphold its mandated responsibilities. Several (male) members of the PMF who had been in less senior positions, and who the City was more inclined to work with, used this email to claim that the women had resigned from the PMF. They got legal support to change the non-profit organisation information related to office-bearers (PMF 2017; PMF women representatives 2017). The original mediator of the Accord (Mediator 1 2017) reinforced this view explaining that:

City officials decided to collapse the leadership of the PMF and install a puppet leadership. Elite capture was then permitted (a group of males within the leadership were supported). These males were more amenable to being the agents of the City and Province instead of the representatives of the community and partners to the peace process agreements. The unlawful and unethical action created contradictions and new conflicts within the leadership of the PMF.

Subsequent to this there have been complaints from others in Hangberg (Pastor 1 2015; Hangberg women's network leader 2015) that PMF elections are held in small venues, thus many residents feel excluded from discussions as they cannot see what is taking place from outside, or they implicitly feel unwanted. An HBCA member describes how:

> Those ... people only talk to themselves, they have private meetings and they decide there what they gonna do for the whole of Hout Bay and without us knowing, we just read it in the papers maybe a day or two before something starts. So we don't have time to query or ask questions, the decision is already made and yet the mayor, city council, they meet every month with PMF ... you can ask anyone, no public meetings, only meetings with PMF leadership. They are the only people that know what's going on, what's going down with this. It's closed, ja.
>
> (HBCA member 1 2015)

Co-governance is only an effective democratic and developmental tool when societal actors are genuinely able to hold bureaucrats (and elected leaders) to account. In the case of the PMF it seems that this did not happen. Instead, 'leaders' more amenable to co-option replaced rather those that challenged the City and Accord partners. These new leaders also now face the challenge of legitimacy from below.

State dependency and patronage

The limited capacity of the PMF to co-govern developmental governance in Hangberg means that the initiative for designing and implementing the projects listed in the 2011 Accord rests with the various branches of the state. As described in Chapter 2, the state works in rule-bound, hierarchical silos that make it systematic but slow, narrow in scope, but focused. In addition, the PMF must engage separate spheres of the state as well as different departments, adding significant complexity (and thus time) to the project process. Without real capacity to hold the state accountable, progress on projects in Hangberg ultimately rests on the discretion of these different branches of the state. They set and drive the agenda for change, with the PMF's role effectively limited to forms of consultation. The internal divisions in the PMF noted above, also due to poor organisational design, have served to slow the pace of delivery further. Thus, poor institutional design leads to the weak legitimacy of PMF leaders and their dependency on state initiative. Together these factors slow, rather than speed up, development.

In addition, however, an important further consequence of the Accord is that the space for independent civil society in developmental governance has been closed down. According to a current PMF leader,

> the Forum is actually the only recognised legal entity within the community that has a right to liaise with City, SANParks and Province. So whenever there are jobs or anything that is coming out regarding the community, the Hangberg community have to speak only to the Peace and Mediation Forum.
>
> (PMF leader B 2015)

The head of Informal Settlement Upgrading for the City (City official 2 2015) agrees that 'although there are a number of structures ... we only work with the PMF'. The fact that the state is legally bound to work with one structure, albeit one that is democratically elected, can easily lead to the creation of patronage networks.

Clientelism and patronage are contested concepts. Earlier writing on clientelism frequently viewed the clientelistic relationship as one of domination and inequality where the patron dominated the client, including different forms of social interaction such as exchange, conflict, and prostitution. Clientelistic relations constituted a realm of submission; a pillar of oligarchic domination that reinforced and perpetuated the role of traditional political elites (Hagopian in Auyero 1999). Focus has often been placed on how political parties use material incentives to win votes, but do little to change the lives of voters beyond this. However, Fox (2012) explains that this is too exclusive a view as it does not take into account the various ways in which state actors use their control over access to public resources to manipulate citizens. Indeed, brokers need not always be politicians. Clientelism can also be experienced in both individual and collective settings. Based on research in Brazil, Gay (1998: 14) argues that clientelism is increasingly a means to pursue the delivery of collective interests as opposed to individual goods. Political clients are therefore more likely to assume the form of organisations and communities that 'fashion relationships or reach understandings with politicians, public officials and administrations'.

Indeed, there is evidence of a perception that leaders of the PMF, and others with whom they associate closely, benefit unduly from their access to the state. They are perceived to allocate resources in a biased manner, based on their social networks. This view was expressed by some respondents from Hangberg but also by community leaders from neighbouring Imizamo Yethu (ANC leader 2015). The capture of opportunities for personal networks was a problem also associated with HiDA. As one resident of the Hangberg community described it, 'there [are] always projects coming up like for computer courses, and you will only see the same people attending these and we know it's family because everyone knows each other ... it's the same people year in, year out ... they are the ones the information comes to, which they keep in the family' (Waste activist 2015). A previous member of the PMF stopped working with them out of frustration over the fact that 'they work for themselves', and that they decide for, rather than with, the community about the allocation of resources. Hangberg youth in a focus group also stated that 'those people [PMF leaders] are naaiers [bad people] ... who just cater after themselves ... the people here don't have jobs ... we don't have jobs ... but they look after themselves and their friends and family'.

An HBCA member believes that the PMF do not listen to, or help, members of the ANC or its aligned HBCA but, 'their people, like their immediate family, their party members get sorted out fully' (HBCA member 1 2015). According to an HBCA member,

> if the Hout Bay Civic Association asks for help or complains about problems or wanting to raise issues since 2010 – nothing – they don't what to listen to the civic association anymore, they don't wanna do anything, they just work with the PMF.
>
> (HBCA member 1 2015)

These viewpoints are disputed by the PMF, who argue that PMF leaders do quite the opposite: they spend a lot of personal time and expense on solving residents' daily (and longer-term) problems without financial reward. A PMF leader (PMF leader B 2015) explains that the PMF only has a small advice office with no other resources. Indeed this was behind the frustration of the six women who were in conflict with the City. They explained that PMF work was not remunerated, but took a lot of time and personal resources, such as mobile phone airtime.

The implications of PMF capture by a small network, if true, are especially portentous for development in Hangberg, given the PMF's monopoly on access to the state. As a member of the HBCA claims, the PMF hold too much power, 'every organisation, every work that comes to Hangberg is gonna go through that organisation' (HBCA member 1 2015). Thus, the problem is not just that the PMF might have been corrupted to serve the few, but that it also delegitimises other social formations that might want to initiate projects of their own. In this respect, there is ample evidence to show that development projects are often stalled by local politics in Hout Bay. In this regard, the story of land in Imizamo Yethu and the multiple SANCOs discussed in Chapter 6 is a good example. Local community divisions can make it challenging for a government department to complete its work. However, this is a narrow view of the problem as it is often the development projects themselves that lead to conflict. They create a set of winners and losers, insiders and outsiders. Residents feel under-represented and that their interests are not being met, while others appear to benefit unduly. It also leads to clientelistic politics where residents are highly incentivised to gain access to a channel of state resources.

Conclusion: from participation to representation

There are two interesting reflections on democracy and developmental governance emanating from the Battle of Hangberg and the consequent formation and functioning of the PMF. First is that the Battle reflects a failure of formal local democracy to deal with the needs and concerns of Hangberg residents around land and housing insecurity. The local ward councillor did not address these concerns, and in fact exacerbated them with discussions of moving people to 'Blikkiesdorp'. The former ward councillor had a minimal relationship with residents of Hangberg and played almost no role in the UISP project. No councillor's name was mentioned in any interview about the project and no interviewee ever referred to the ward councillor as a point of access to the state or City Council in terms of development or service delivery. The provincial Premier, too, was also unable to facilitate mediation and resorted to the coercive arm of the state rather than an attempt to reach a negotiated settlement (Development practitioner 2017). The weakness of formal democracy, including participatory processes, led to mobilisation and protest by Hangberg residents.

Second, the outcome of mobilisation led to a court-sanctioned 'invited space' to supplant civil society in developmental governance: the PMF. The formal democratic system does not establish political representation at any level below the ward, but as Chapter 6 also demonstrates, all forms of developmental governance require representation at the level of the settlement. A consequence of this (often-contested) representation is the creation of opportunities for gate keeping, brokerage or mediation linked to the distribution of resources. This brokerage can often turn into patronage linked to political party affiliation, in other words, political clientelism.

It was precisely this partisan contestation that contributed to the Battle of Hangberg, and that the PMF was set up to transcend. Thus, the order of the court forced the City, Province, and community leaders into a formal relationship of co-governance. While co-governance has the potential to deepen democracy through new forms of participation and representation, and to improve service delivery, it has not fulfilled that potential in the case of Hangberg. As argued above, the poor institutional design of the PMF means it is practically unable to co-govern with the state, and lacks clear and legitimate rules that constitute its leadership and operation. It certainly cannot implement any decisions reached on services or projects. In a context of partisan and personal contestation, these features have further weakened the PMF, and slowed the implementation of development projects in Hangberg even further. At the same time, other civil society formations are prevented from engaging the state on development projects, given the monopoly of representation established by the courts. Consequently, for residents of Hangberg, not only has participation in developmental governance been transformed into representation via the PMF, but into a monopoly form probably captured by a local network. In effect, then, through the failures of the PMF, participation has turned into patronage.

References

Ackerman, J. (2004). Co-Governance for Accountability: Beyond 'Exit' and 'Voice'. *World Development, 32*(3), 447–463.

Auyero, J. (1999). From the Client's Point(s) of View': How Poor People Perceive and Evaluate Political Clientelism. *Theory and Society, 28*, 297–334.

Bakkes, A. (2011). Fists Fly at Hangberg Meeting. *The People's Post.* https://issuu. com/thepeoplespost/docs/peoples_post_constantia_wynberg_edition_25.10.11. Accessed 8 December 2015.

Bezuidenhout, N., & Williams M. (2014). Cars set alight as Hangberg erupts. *IOL online*, 30 September. www.iol.co.za/news/cars-set-alight-as-hangberg-erupts-1757936. Accessed 2 December 2015.

Buhler, A. J. (2014). Smiling in the face of precarity: housing and eviction in Hangberg, South Africa. Doctoral dissertation, University of Cape Town.

de Greef, K. (2013). The booming illegal abalone fishery in Hangberg: tough lessons for small-scale fishiers governance in South Africa. Master's thesis, University of Cape Town.

De Lille, P. (2011). Hangberg Peace Accord Becomes An Order Of Court. Statement by the Executive Mayor Alderman Patricia De Lille. *City of Cape Town.* Available at: www.capetown.gov.za/en/MediaReleases/Pages/HANGBERGPEACEACCOR DBECOMESANORDEROFCOURT.aspx. Accessed 31 August 2015.

Ehebrecht, D. (2014). The challenge of informal settlement upgrading: breaking new ground in Hangberg Cape Town? Master's thesis, Universitätsverlag Potsdam, Potsdam.

Fieuw, W. V. P. (2011). Informal settlement upgrading in Cape Town's Hangberg: local government, urban governance and the 'right to the city'. Master's thesis, Stellenbosch University.

Fox, J. (2012). State Power and Clientelism: Eight Propositions for Discussion. In Hilgers, T. (Ed.), *Clientelism in Everyday Latin America Politics.* New York: Palgrave Macmillan, 187–213.

Gay, R. (1998). Rethinking Clientelism: Demands, Discourses and Practices in Contemporary Brazil. *European Review of Latin American and Caribbean Studies, 65*, 7–24.

HPMA (Hangberg Peace and Mediation Accord). (2011). Western Cape High Court Case No. 21643/2010. 19 October 2011.

Joubert, P. (2007). DA Rep Calls for Forced Removals. *Mail&Guardian Online*, 16 February. https://mg.co.za/article/2007-02-16-da-rep-calls-for-forced-removals. Accessed 10 December 2015.

Kaganof, A., & Valley, D. (2010). *The Uprising of Hangberg.* www.youtube.com/ watch?v=bRjMB3znA2E. Accessed on 1 May 2015.

Majavu, A. (2010). Bitter gray areas in Hangberg clash. *Sowetan Live*, 30 September. www.sowetanlive.co.za/opinion/columnists/2010-09-30-bitter-gray-areas-in-hangberg-clash/ Accessed 1 December 2015.

Piper, L., & Anciano, F. (2015). Party Over Outsiders, Centre Over Branch: How ANC Dominance Works at the Community Level in South Africa. *Transformation: Criti cal Perspectives on Southern Africa, 87*(1), 72–94.

PMF women representatives. (2017). Email correspondence between PMF women, PMF leaders and City of Cape Town officials. May 2017.

Rubin, M., & Royston, L. (2008). *Scoping Study – Local Land Registration Practices in South Africa, Urban Landmark.* www.urbanlandmark.org.za/downloads/Local_ Land_Registration_Practices.pdf. Accessed on 10 December 2015.

SAPA (South African Press Association). (2010). SAN Parks now a respondent in Hangberg case. *Times Live*, 18 October. www.timeslive.co.za/local/2010/10/18/ sanparks-now-respondent-in-hangberg-case#. Accessed 30 September 2015.

Stone, C. (1989). *Regime Politics: Governing Atlanta, 1946–1988.* Lawrence, KS: University Press of Kansas.

Stone, C. (2006). Power, Reform, and Urban Regime Analysis. *City & Community*, 5(1), 23–38.

Zille, H. (2014). SA Today: The Salutary Story of Hangberg. *Bokamoso News*. www.da.org.za/2014/03/sa-today-the-salutary-story-of-hangberg/. Accessed 15 July 2015.

Interviews

1. Air pollution activist. (2017). Interviewed by Fiona Anciano. 23 August 2017.
2. ANC leader. (2015). Interviewed by Laurence Piper. 20 May 2015.
3. City official 2. (2015). Interviewed by Fiona Anciano. 30 March 2015.
4. SSP patroller. (2016). Interviewed by Fiona Anciano. 23 November 2016.
5. Development practitioner. (2017). Interviewed by Fiona Anciano. 29 November 2017.
6. Hangberg women's network leader. (2015). Interviewed by Fiona Anciano. 17 August 2015.
7. HBCA member 1. (2015). Interviewed by Fiona Anciano. 6 March 2015.
8. Khoisan Chief. (2015). Interviewed by Zikhona Sikota. 6 March 2015.
9. Mediator 1. (2017). Email correspondence between Mediator and Fiona Anciano and Laurence Piper. 17 August 2017.
10. Pastor 1. (2015). Interviewed by Fiona Anciano. 17 August 2015.
11. PMF. (2017). Focus group with first leaders of Hangberg Peace and Mediation Forum. Interviewed by Fiona Anciano and Laurence Piper. 24 August 2017.
12. PMF leader A. (2015). Interviewed by Fiona Anciano. 17 March 2015.
13. PMF leader B. (2015). Interviewed by Fiona Anciano. 20 February 2015.
14. Political representative 1. (2017). Interviewed by Fiona Anciano. 21 February 2017.
15. Waste activist. (2015). Interviewed by Laurence Piper and Fiona Anciano. 20 February 2015.
16. Western Cape Provincial government. (2017). Email correspondence with Helen Zille, Anthony Hazell, Laurence Piper and Fiona Anciano. 8 November 2017.

5 Poaching the bay

Turning fisherfolk into smugglers

> Poaching is how Hangberg survives. Most of us don't have jobs or a good
> education. We poach and hustle to keep our mouths and our children's
> mouths open.
>
> (Isaacs 2012)

Taped to the wall of a small concrete house perched on the side of Sentinel
Mountain are nearly a dozen individual pictures of men, all previous residents
of Hangberg. These, an interviewee says, are the photographs of his friends
who died while diving off the coastline surrounding Hout Bay. As the conversation progresses, it becomes clear that over the past 15 years there have
been many deaths of Hangberg residents due to late night poaching in the
rough waters around Hout Bay. These deaths are not isolated or infrequent.
An Internet search brings up dozens of newspaper articles and stories on such
drownings.

There is no fixed number of deaths because some are due to illegal activities
and so are not reported. For that matter, not all poaching deaths are due to
drowning. Families of the deceased also claim there are incidents where rival
poachers kill their competitors at night and throw their bodies into the sea
(Joseph 2001). Although difficult to quantify – one article claims 13 men died
over an eight-month period in 2006 alone (Dolley 2006) – there are undoubtedly serious and deep-rooted concerns around livelihoods and illegality in
Hangberg. Indeed, as this interviewee looks at the pictures of his friends who
have died, he explains, 'What they were doing was illegal, but it was the only
way they could help to feed their family' (Fisher 1 2015).

Why has poaching become a way of life for so many residents in Hangberg?
What are the underlying political, democratic and governance issues driving
this move to illegality? This chapter will chart the story of informality in
Hout Bay through an analysis of the dominant forms of governance affecting
small-scale fishers in the community. It will start with a discussion of the evolution of small-scale fishing policy, showing how this form of developmental
governance has ultimately denied long-standing fishers the right to fish. It will
then demonstrate how fishers experience weak representative and participatory democracy but strong market governance, and how this combination has

Figure 5.1 Poachers entering the sea in Hangberg

led to rising informality and even a politics of criminalisation from the state and society. In essence, it will show how weak state rule and strong market governance produce the conditions that make informal and illegal livelihoods inevitable.

Fishing in Hout Bay

Hangberg has a long history as a small-scale fishing community. Traditional fishing practices in fact predate the formal establishment of Hangberg as a residential area. Between the 1950s and 1980s, industrial fishing and pro-cessing activities formed the foundation of the local economy in Hangberg. A large proportion of the population worked in the fish factories near the harbour and on industrial fishing vessels (Schultz 2015). History shows that fishing, particularly for rock lobster and abalone,[1] is woven into the social fabric of the Hangberg community. Many of the current informal boat owners share stories of growing up on their fathers' fishing boats (Hauck 2009). In a focus group conducted with poachers in Hangberg in early 2016, it was clear that they all came from families steeped in the history of fishing off the Hout

Bay coast (Poachers 2016). As one interviewee explained, 'I'm on the sea for more than 20 years. I grew up on the seas. I'm a fisherman all the years ... my father's been a fisherman ... my mother worked in the processing plants' (Fisher 2 2016).

These historical fishing practices, and the economic livelihoods of Hangberg residents, were badly affected by government policy on fishing quotas. Traditional fishing practices are now subject to modern laws and policies. Indeed, as recently as September 2017, large-scale protests erupted in Hangberg over the Department of Agriculture, Forestry and Fisheries' decision to cut fishing quotas (Mortlock 2017). Clearly, governments play an important role in regulating the limits of harvesting marine life. Sustainable fishing relies on the ability of fish, including abalone and crayfish, to replenish themselves through natural reproduction. If too few fish are left in the sea, the fish population is unable to recover. This is the biological meaning of 'overfishing' (Allison 2001: 934–935). Thus, a strong and effective management strategy by state institutions is essential to prevent the temptation by seaside communities to overfish.

Fisheries policy

Over the years, South Africa has tried to implement several fishing policies related to small-scale fishers such as those living in Hangberg. During apartheid, a limited number of white, South African commercial firms dominated the fisheries sector. Thus, one task of the government in 1994 was to initiate the transformation of the fishing industry through implementing a fisheries policy that would integrate historically disadvantaged black South Africans into the sector. In this context, the Marine Living Resources Act (MLRA) was passed in 1998 to foster equity, the sustainability and the stability of the fishing industry (Stern 2013). Although various fisheries reforms have tried to extend access rights to traditional fishers, these processes have largely reinforced the status quo, leaving big capital to monopolise fishing rights and turning traditional fishers into illegal fishers (Hauck 2009). Indeed, researchers argue that neo-liberal approaches to economic policy in South Africa have influenced fisheries governance, leading to the prioritisation of profitable, capital-intensive, export-driven industrial production, at the expense of sustainability, social justice, and equity (Schultz 2015).

Ongoing concerns about fisheries policy led to legal action against the state by traditional fishers, supported by the Human Rights Commission. The fishers' action against the Department of Environmental Affairs and Tourism (DEAT), (now the Department of Agriculture, Fisheries and Forestry (DAFF)) resulted in an out of court settlement that required the Minister to initiate a new policy process (Hauck 2009). DEAT subsequently mandated a national task team, that included representatives of fishing communities, to craft a new Small-scale Fishing (SSF 2012) policy. This policy moves markedly away from the commercial sectoral focus of earlier fishing policies (Masifundise Development Trust et al. 2012). The Policy for the small-scale fisheries sector became law in June 2012. A central objective of the new SSF policy is to recognise the socio-cultural-ecological and economic linkages of small-scale fishery systems. This approach, in theory, promotes the sustainable use and management of marine resources while having a human rights focus and prioritising socio-economic development, food security and poverty alleviation (Masifundise Development Trust et al. 2012). The policy is targeted at small-scale fishing communities and small-scale fishers whose livelihoods and socio-cultural identities rely on marine resources and who exploit these resources with little to no technology. The policy supports the allocation of a long-term, multi-species right to source marine resources close to the shore (Masifundise Development Trust et al. 2012).

An important aspect of the SSF policy is the stipulation that fishers must now form community-based legal entities (CBLEs) to claims rights for their communities (Stern 2013). The prerogative to issue these rights lies with the Minister of DAFF. CBLEs will be part of co-management structures, a formalised structure that serves as the main local management responsible to oversee compliance, protect community interests, and ensure sustainability and conservation of ecosystems and marine resources. A fisher can only become a member if they are an unemployed South African and can demonstrate daily or historical involvement in the fishing value chain (harvesting, processing, and marketing) over a period of 10 years, with historical dependence on fishing, proof of subsistence on their catch, and/or involvement in semi-commercial activity, like barter or sale. These individuals will be identified by the community and placed on a list verified by the Minister (or the Department or a third party), and then become a CBLE (Masifundise Development Trust et al. 2012).[2]

In practice, however, the MLRA and SSF policy have been unable to deal with growing livelihood insecurities and have not transitioned illegal fishers into legal subsistence fishers. Why have the policies been unsuccessful? The allocation of fishing rights is perhaps the most contentious issue in fisheries governance. Schultz (2015) argues in his research, and indeed our research confirms his view, that a majority of small-scale fishers feel that fishing rights have not been granted in an equitable manner. The total allowable catch (TAC) favours industrial companies, leaving individuals with a disproportionally small part of the total allocation. This inequity is felt at the local level too, with the question of who is allocated rights in a community such as Hangberg, causing tension.

Instead of targeting historical fishers, the initial permit reallocation that was implemented to achieve the MLRA goals was opened to all. This resulted in many permits being allocated to new entrants (including white commercial players), who were able to use skills and business acumen to apply for permits rather than the intended beneficiaries (Stern 2013). Even where permits were allocated to historical fishers, fishers have indicated that the allocated quotas are too small. It is common in many countries to find that prescribed quotas are considerably lower than what individual fishers would choose for themselves (Kuperan & Sutinen 1998). Once fishing quotas are in place, however, anyone fishing without a licence or exceeding their quota is engaging in an illegal activity.

A further concern with fishing rights is linked to the verification of permit allocations. Fishers have to be seen as 'bona fide' in their applications and some local fisher representatives may have exploited this verification process by facilitating access to permits on condition that they are granted authority to market the permit holder's catch for a percentage of the sale price. This process leads to indiscriminate endorsement of as many permit holders as possible. DAFF has also been unable to manage the verification process effectively, leading to the inequitable pattern of fishing rights allocation processes being repeated (Schultz 2015). A related concern is that the communities affected by fishing policies have no means to monitor the application of the permit allocation, or stop outsiders from illegally fishing in the waters near their communities. The SSF policy tries to deal with this by allowing for the training and appointment of a community monitor (albeit with unclear powers of arrest or punishment), and with the allocation of fishing rights to communities, not individuals (Stern 2013).

Although the SSF policy is a step in the right direction in supporting small-scale fishers, it is clear that the vast majority of fishers in a community such as Hangberg are still conducting illegal activities when they fish. An informal fisher, whose quota application was denied on many occasions, explained that 'it is the government who has made us illegal. We have been fishing [lobster] all our lives and it is their problem that they see it as a crime. We just see it as a way to make a living' (cited in Hauck 2009: 155). Indeed, in a focus group with poachers from Hangberg, they were clear that the SSF policy has not worked for them. They do not want to be seen as 'small-scale'. They ask why they are not supported to become commercial fishers. They have developed comprehensive plans to open an abalone farming business but, after many years of engaging with DAFF, they have still not been issued a licence. The group have also spent over R30,000 establishing a formal co-operative. The application for their co-operative to receive fishing rights was sent to DAFF 4 years ago and they are still waiting for a response (Poachers 2016).

The fishers were also unhappy that the SSF policy does not issue individual rights. Large collective groups are not easy to manage and are likely to lead to inter-community tensions. The group has a clear memory of co-operatives that were formed in the 1990s with established commercial fishers. They describe how the ordinary members were marginalised from profits and eventually saw no benefit from the venture while two or three commercial individuals benefitted. More established fishers, who have had formal fishing rights allocated or been involved in commercial fishing, are also unhappy with the structure of the SSF policy (Fishers 3 & 4 2016). They describe how tensions invariably result over the formation of co-operatives in Hangberg. With the best of intentions, old friends will fight over who has which position in the co-operative and the allocation of resources. For fishers in Hangberg, small-scale fishing policy has created a sense of frustration and marginalisation.

Democracy, parties, and poaching

If small-scale fishers feel state policy on fishing rights and allocation is problematic, how then, if at all, have they tried to influence this policy? Remarkably, fishers in Hangberg have been surprisingly active in their attempts to engage the state. A vocal minority have tried numerous avenues to influence fisheries policy, government officials, and politicians. Illegal fishers hoped to influence decision-making and policy around their fishing rights by accessing the formal representative democratic system, and by using participatory governance through engaging with the multidimensional state. These attempts at engaging the democratic system, as we shall demonstrate, have not been effective.

The fishers we spoke to in Hout Bay, both legal and illegal, have tried numerous ways to engage with the formal democratic system. Many of them, at different times in the past few decades, have joined or supported political parties. Originally, respondents felt the ANC was their main hope of support, especially after the democratic transition in 1994 when the ANC wielded widespread power nationally and provincially (Poachers 2016; Fisher 1 2015). As the voting patterns in Chapter 1 demonstrate, the ANC received the highest number of votes in Hangberg in elections after 1994, before this pattern changed in the mid-2000s. As a member of the focus group explains 'the people' of Hangberg tried to access opportunities through political parties 'back in the day … but it has stopped'.

Initially, poachers believed that being affiliated to a political party that was in power would increase their chances of receiving fishing permits or influencing policymaking. In the late 1990s, they started to work with the ANC through one of its local leaders. As they explain:

> All the things we did for the ANC … we done everything … so we decided since our story is taking so long that we are gonna hook up with [a local ANC leader], intermingle with him … through this intermingling we had to do a lot of things for the ANC … canvassing and getting votes for the ANC. … I mean we did a lot of things for the ANC … no result! No result!
>
> (Poachers 2016)

Other interviewees working in the fishing industry also explain how they were told to 'hook-up with the ANC' if they wanted to improve their chances of getting better fishing quotas (Fisher 1 2015).

A more recent attempt to use politics to win influence can be seen through the harbour project in Hout Bay. Hout Bay has an active working harbour that is used by commercial fishers, small-scale fishers, tourist boats, and poachers. Over time, a gentrification process has taken place in the harbour, with the main activities moving from fishing to tourism. This has prompted a rethink of the harbour design and use. Harbours are national mandates, controlled by DAFF and the Department of Public Works. Due to their large economic potential, the DA provincial government in the Western Cape is very keen take control of the harbour (Zille 2017). For this reason, one respondent in our poaching focus group, who currently supports the ANC, explained:

> I've seen DA's policies and what Helen Zille said ... they want to control the harbours, because it is currently controlled by national government ... and she's always bad mouthing the harbours for its sunken ships, 'lawlessness' and poaching, etc. They use Hout Bay for a political football ... they are all after the harbours and don't make a secret of it.
>
> (Poachers 2016)

The same focus group respondent has a close family relative who sits on the DA-aligned PMF, discussed in Chapter 4. He feels, however, that working with the ANC may be more beneficial in the current political climate, as it may improve his ability to benefit from the harbour redesign. Indeed, he is a representative of the ANC-aligned HBCA. A colleague of his who works with the respondent on an aqua farming project was very surprised to hear that he had joined the HBCA. He explained that this was 'news to me' but that it makes perfect sense politically:

> You must remember national government runs our natural resources, our seas, our fishing quotas and the minister comes from the [ANC-aligned] SACP [South African Communist Party] ... And his deputy Bheki Cele is from the ANC and they have a very specific mandate and that is transformation in the fishing industry. Now these guys didn't get a quota, he's a poacher ... in other words ... he decided to jump ship to the ANC with the hope that the ANC will give him a quota that's coming up in 2016. With the hope that with the ANC will come money and help to build this [aqua] farm.
>
> (PMF leader A 2015)

The flexible support for political parties that may assist in improving personal livelihoods is most starkly evident in a statement from an ageing Rasta poacher (Poachers 2016). He describes with emotion how, even after he has suffered under apartheid, he would vote for any party that would support his ability to become a legitimate fisher, even if that party was conservative and white-led:

> If any party can come and say this is what we can give you, we just need to secure your vote ... and I can see what they are offering is real and true.

... I'll vote for them yes ... a hundred times ... the EFF ... even for the Freedom Front Plus!![3]

It is clear that the fishers we spoke to saw a dual benefit of supporting a political party. First, if the party won power (either local, provincial, or national), there was a hope that political representatives would use formal democratic channels to support the plight of illegal fishers through changing fishing policy, which would lead to the allocation of larger quotas for small-scale fishers. This faith in representative democracy was particularly relevant in the first decade after 1994.

The second benefit of supporting a political party is more closely related to the idea of clientelism than the idea of representative democracy. Here respondents describe how they were promised direct benefits if they 'worked for' the ANC. When discussing how quotas are allocated and who ends up running co-operatives, one fisher explained that it's all about who you know that will determine if you get in.

> At the beginning they were 10 clever guys, these directors, who were not fishermen. They had all these connections with ANC officials ... and they applied for the quota. However, they were turned down and told that they need 20 fishermen to succeed with their application. Then they came to approach us 20 fishermen.
>
> (Fisher 5 2016)

Although the co-operative was successfully set up, over time the non-politically connected fishers were eased out of their role in the co-operative. Most signed their shares away, or remain in the co-operative on paper only, receiving few, if any, dividends (Fisher 5 2016).

Fishers, however, have tried to engage with the political system well beyond elections and supporting political parties. They have formed groups to represent them in engagements with multiple levels of the state. The attempt by fishers in Hangberg to engage in participatory governance practices must be seen within the context of a national fisheries governance policy that has increasingly promoted equitable and inclusive engagement between state and non-state actors. Small-scale fishing communities are now seen as partners who should engage with state management officials, fisheries scientists, and fishing company directors on equal terms (Schultz 2015; Masifundise Development Trust et al. 2012). As Schultz (2015: 33) explains, in South Africa 'political representation constitutes a primary mechanism for facilitating the democratic engagement of small-scale fishing communities in fisheries policy and management processes'. Fishers in Hangberg, therefore, have tried to influence the multiple actors of the state at all levels.

These attempts at representation and participation have been through both 'invited' and 'invented' spaces (Cornwall 2002). In terms of invited spaces, government officials were adamant that small-scale fishers be involved in the drafting of the SSF policy, with general recognition that previous efforts to draft a policy for small-scale fishers had failed because it was top-down and did not have community input and ownership. Approximately 80 fisher representatives (15 to 20 fisher representatives per province) were invited to a summit for a stakeholder engagement process to guide the development of this policy in 2007 (McDaid 2014). The fishers we interviewed in Hangberg referred to these representative groups, explaining how there were different groups that 'helped us to fight' (Poachers 2016). Although they were optimistic about these representative forums at first, they soon became disillusioned:

> We are on our own ... in the past organisations used to represent us ... then we tell them what to say in these meetings and they don't say the things we mandated them to say ... all those other people who represented the fishing communities that helped us to fight but didn't represent us accurately ... they push their own agenda. We decided to do our own thing.
>
> (Poachers 2016)

The poachers explained that they have now formed their own association to represent those who want to move from poaching into abalone farming, calling themselves the Hout Bay Aqua Farmers. Yet there is frustration here, too, with a lack of progress in influencing government policy on aqua farming or in generating support for abalone farming: 'No brother ... we don't even get recognised ... it's [messed] up man ... we even travelled up the coast to various coastal fishing communities like Mosselbay, Struis bay, Lamberts bay, etc. All of them have the same problem'. As one poacher (Fisher 1 2015) explained in relation to his attempt to influence government to support aqua farming, 'Myself have sunk a total of R40,000 on meetings and travelling, phone calls and emailing and stuff. But I won't get tired because this is a big thing.'

The respondents told numerous stories of how they have approached politicians and officials over the years. They described meeting senior politicians, including national and provincial Ministers, in different forums, at different times:

> The team we had always met officials who always sent us around to different officials. ... We called Bheki Cele [then Deputy Minister of DAFF] ... many times and he only came once ... and he promised he'll speak to the Minister ... we waiting until today now ... it's more than 4 years. ... Rob Davies [Minister of Trade and Industry] we spoke to him after he signed off that interim relief and boats ... we spoke to thousands of officials ... Treasury, DTI, Public Works.
>
> (Poachers 2016)

Neither formal democratic engagement with political parties and voting, nor participatory governance through forums and meetings, has led to any noticeable accountability or responsiveness in terms of small-scale fisheries governance, according to those we spoke to in Hangberg. All interviewees were quite clear that the democratic system was not benefitting them as a group or personally. They were clear that supporting a political party would not create better representation for their community: 'Voting for the ANC wasn't worth the time ... to be honest I am not interested in any political parties' (Fisher 1 2015). As a member of the focus group explained in reference to participatory governance, 'these guys made a lot of promises and more ... we've been to many places and shown our plans and had meetings ... but nothing brother' and another noted that, 'We've been through a lot of ministers. As I see it the government simply don't want to help the people of Hangberg' (Poachers 2016).

For researchers who look at fisheries governance there is a fundamental tension between democratic reform of fisheries policy, and the profound structural inequalities of power that shape how fisheries governance unfolds in practice. They argue that the failure to locate power relations associated with economic exploitation and appropriation results in governance approaches that are not empowering for small-scale fisheries (Schultz 2015; Davis & Ruddle 2012). It is clear from our research that fishers feel there is ongoing inequitable distribution of fishing permits, and that they are unable to obtain abalone farming rights democratically. They also feel strongly that no party represents their interests and that the democratic system itself is not effective. In this sense, they are profoundly disconnected from democracy.

Market governance: from fishers to poachers

If neither representative nor participatory democracy is a form of governance that affects the everyday realities of poachers in Hangberg, then what form of governance does affect them? It becomes clear after spending time with poachers and legal fishers in Hangberg that it is the broader economic market that influences many of their day-to-day decisions. Fishers in Hangberg are, in a sense, governed by the market and developmental governance in the form of national policy far more than they are governed by the local democratic system.

The market influences small-scale fishers in several ways. First, the market affects how those who do receive fishing allocations operate, and second, a criminal market for abalone and crayfish creates an incentive to turn towards illegal fishing. There are fishers in Hangberg who have received permit allocations for a range of species over the years. While fishing quotas are hard to come by, those who do get them, should, in the framework of fishing policy, be able to develop sustainable livelihoods. Stories from fishers in Hangberg, however, demonstrate that this is not the case. Ultimately, market conditions are not conducive to small-scale fishers developing sustainable businesses due to the economies of scale. Key here is that large commercial players dominate the fishing market in South Africa. According to respondents (Fishers 3 & 4

2016), there are five such 'big players' and small fishers cannot compete with the scale of their operations and their ability to control the market.

It is for this reason that the SSF policy supports the formation of co-operatives. However, as discussed, these structures have not been successful in the past in Hangberg, largely due to infighting and divisions among co-operative members (Poachers 2016). The current CBLE policy within the SSF policy is not yet operationalised in Hangberg; those who have formed CBLEs are still awaiting communication from DAFF (Poachers 2016). There is perhaps more optimism in terms of success as the policy supports the participation of communities in the entire value chain, including harvesting, processing, and marketing. The government will, in theory, provide CBLEs with development finance and training to do their own storage, transport, packaging, marketing, management, finance, logistics, and human resource through locally community-owned and -based marketing firms (Masifundise Development Trust et al. 2012). Yet, there is no evidence of this occurring in Hangberg, even 5 years after the formation of the policy: 'It's not happening, not yet. I don't know where the funds would come from ... that's why I left the industry ... the bureaucracy involved and the turnaround time ... makes people frustrated' (Fishers 3 & 4 2016).

A second problem facing small-scale fishers is the inherent structural inequalities in the market, stemming from the racial legacy of apartheid. Many of the successful fishing companies were started under apartheid and it is therefore often white-run commercial ventures that have the capital to finance the infrastructure, such as boats, required to maintain a successful business. Allied to this historical monopoly in the market is the reality that black small-scale fishers who are awarded quotas often lack financial expertise or experience. As fishers from Hangberg explained,

> historically disadvantaged people ... don't understand the monetary value of their rights ... if you never saw a R100,000 or R80,000 in your account ... let's be honest ... there was no financial planning ... someone explaining to you that you have to pay tax ... the intention of the policy was right but there were no support structures for the people. ... I think that's why they lost their rights.
>
> (Fishers 3 & 4 2016)

Fishers explain how they sold their quota rights to 'the old role players' or new elite-controlled companies to generate immediate income (Fisher 5 2016). Even where fishers do not sell their rights it is very difficult for individuals, or even collectives, to get bank loans to buy boats and other equipment. Thus, small-scale fishers are often compelled to work with more established fishing companies, resulting in lost income:

> They will tell you that you can get the quota because you are black but you need to come to me because you don't have a boat. That's when the

problem of the middle man comes in where instead of giving you R180 a tray he will give you R120 a tray then sell it for R400 ... it is very difficult for us ... I have 800kg [quota] and say collectively we have 2 tons ... what you gonna do with that ... that's 1.5 million rand ... and the bank will just laugh at you. They won't give you a loan. You see we got all these nice policies in place but the process, at the end of the day you just give up.

(Fishers 3 & 4 2016)

That market governance affecting everyday realities and structural poverty in Hangberg is clearly explained by fisheries researchers Davis and Ruddle (2012: 249):

Marine harvesters must engage in exchange relationships. This means full-time fishing demands the production of commodities for exchange or sale, and it follows that the material quality of harvesters' lives depends on the terms of economic exchange values. That is, harvesters are impoverished by political and economic circumstances they generally cannot control. Thus, the local, regional, national, and international political economy of commodity values, wealth distribution and accumulation, power, and class are ... germane to understanding material poverty.

Poaching and market governance

Given the challenges of working in legal markets for those who are lucky enough to get quotas, it is not surprising that so many small-scale fishers turn to poaching to make a living. Illegal fishing is often the only option of generating an income for those who do not have fishing quotas. These fishers are then compelled to operate in the illegal market when they want to sell their trade. If fishers harvest their catch without a permit, there is no legal way for them to sell their catch. Rock lobster and abalone are the two main sources of illegal fishing in Hangberg, although lobster is less frequently poached nowadays (de Greef 2013; Schultz 2015). Selling these commodities on the open market is difficult for fishers, and an extensive illegal market has thus developed.

Even if fishers have other livelihood strategies, the income they can gain from the illegal market makes poaching an appealing choice. Both lobster and abalone are highly valued shellfish. Abalone, in particular, is a high-value resource for organised illegal fishing networks, especially from buyers in China (Stern 2013). Dried abalone can fetch anything between R6,000 and R12,000 per kg once it reaches Asia (Goga 2014). Fishers and their support crews in Hangberg do not see this scale of income; however, poaching is still lucrative for individuals. A diver will earn approximately R200 to R250/kg of abalone, once sold to a local intermediary (Fishers 3 & 4 2016). The diver will then pay spotters, carriers[4] and other boat crew a share of this income. Boat owners will also get a percentage. Divers can earn approximately R10,000 per operation, while boat owners can earn R7,000. On a good night a carrier can earn up to R1,000. Poachers may have up to twelve operations a month, but generally four or five operations are sufficient to live off (de Greef 2013; Lambrechts & Goga 2016). In comparison, the average salary for a worker on an abalone farm is about R4,000/month (Viceland 2016). Fishers can make considerably more money from poaching than from legal activities such as abalone farming.

Once harvested, poachers sell their catch to middlemen, who then sell on to larger syndicates (de Greef 2013). It is not clear who exactly runs the syndicates that buy abalone from Hangberg poachers, through middlemen. The poachers themselves explain that the abalone is sold to 'the Chinese' and that it is Chinese triads that control the market (Poachers 2016). There is also a perception in law enforcement circles that there are a few big players who have market dominance due the scale of their operations. This view is disputed with evidence showing that the market is neither hierarchical nor particularly concentrated, as multiple players have resources to buy, dry (or can), and then ship the abalone (Steinberg 2005). While the number of players may be disputed, there is an undoubted link between gangs (including triads) and poaching in Hangberg. The abalone trade is now dominated by 'outside opportunists who establish a black market trade for economic gain' (Hauck 2009: 196).

There are numerous social and economic consequences of the effect of black market governance on residents in Hangberg. The 'outside opportunists' have arguably had an impact on gangsterism and drug use in Hangberg. Hence, there is an established link between drug dealers and Asian abalone smugglers in Cape Town. Mandrax or methamphetamines are traded for abalone (Goga 2014). The extent to which drugs (and other associated illegal activities) enter Hangberg through poaching is hard to establish, however. Based on time spent in the Hangberg community, it is clear marijuana is used extensively; however, this is arguably a long-standing social practice and not necessarily linked to crime syndicates. Poachers are also clear that there is no serious gangsterism in the community. Even where appearances may reference gang culture, they explain that they are not gangsters. Gangsterism was 'chased out' of the community in the 2000s (Poachers 2016).

While there may not be established gangs operating in Hangberg, such as those that flourish on the nearby Cape Flats, evidence suggests that poaching has led to changing social structures in the community. This is particularly evident in relation to children who have left school in order to poach, or act as crew and carriers (Community Policing Forum (CPF) member 2016). These children and teenagers are able to make money, and are therefore able to afford alcohol and drugs. Boys as young as 10 are recruited by poaching syndicates to act as lookouts; they learn the trade and soon become divers, immersed in the trade themselves. These boys eschew more insidious gangsterism (which is more dangerous and less rewarding) for poaching, but in the process often drop out of school. They lose interest in school and in any event, after poaching in the evening, are too tired to learn. There is a strong perception that poaching has led to a general loss of values among youth in Hangberg. The ability to get 'easy money' has led to unsustainable patterns of consumption and a disregard for the value of education (Goga 2014; de Greef 2013).

The need, and call, for the state to assist with fixing social problems in Hangberg is expressed clearly by the PMF (as referred to in the previous chapter). The PMF, as a formal court-sanctioned representative of the Hangberg community, stipulated in their 2010 accord that, along with pro- viding housing, the state must also provide support and infrastructure for social services, such as youth programmes, and substance abuse programmes. This is a clear bone of contention between the PMF and the City: the flats have been built, but there is no social support programme. Frustration at this element of the Accord not being met was voiced not only by PMF leaders (PMF leader A 2015), but interestingly also by poachers. A poacher explained that the PMF had failed because it had not delivered the promised social programmes from the City (Poachers 2016). While the failure of the PMF to tackle social concerns is a further indictment on participatory governance, it does highlight the power of market governance. Democratic and participa- tory structures are unable to deal with the effect of the market impact, both on legal fishers who are unable to build viable businesses, and on illegal fishers who have turned to poaching to survive.

Marginalisation and the rise of informality

The impact of developmental governance through fisheries policy coupled with weak representative and participatory governance and strong market governance, has led to a rise in informality in Hangberg. Informality is seen by many authorities, starting with the United Nations, as the 'unregulated, uncontrolled, messy and inefficient settlement and use of land' (Porter 2011: 116). In our case, we could enlist those adjectives to frame how South African authorities see poaching: as the unregulated, uncontrolled, messy, and inefficient extraction and retailing of marine resources. Informality is the other side of planning's ordered, contained, and controlled spaces, and thus creates a policy problem. What this study of poaching highlights, however, is that informality is not unrelated to the formal system. This research demonstrates, as have other studies (Porter 2011; Watson 2011; te Lintelo 2016), that informality is often produced by formal authority such as state departments. In Hangberg, the prevalence of poaching is a direct outcome of fisheries policy in that it declares a long-standing practice as now illegal, and makes it nearly impossible for traditional fishers to con- tinue traditional practices within the new legal framework. Faced with an impossible choice, many choose to risk the law and criminalisation rather than risk starvation.

Figure 5.2 Fisher shacks in Hangberg

Fishers in Hangberg become criminal 'others' outside of, and partly abandoned by, the formal system. It is also important to note that fishers can be trapped in a cycle of illegality. As respondents explained, under earlier fishing policies such as the MRLA (Marine Living Resources Act), once you have a criminal record you are not eligible to apply for a fishing quota (Poachers 2016). Following Chatterjee's (2004) argument, poachers are seen as outside the law and thus 'civil society', and at best become populations to be governed rather than citizens bearing rights. The unintended consequence of developmental state policy has its own further effects: first, the perceptions of disorder and of 'criminals' residing in Hangberg create new spatial and racial segregation, and further entrench that which already exists. Second, those that are 'informalised' take one step further towards exiting the state altogether; and third, informality can lead to resistance and conflict. All these consequences of informality make informal fishers, as urban residents, feel progressively more marginalised from the state and the broader community in which they live.

There is a widespread narrative among many coloured residents in South Africa that during apartheid they were 'not white enough'; now, under a black-led ANC government that pursues race-based affirmative action policies, they 'are not black enough' (Anciano 2014: 18). Poachers clearly voice this sense of perceived marginalisation when they explain that, 'we were the ones who fought apartheid with our brothers ... now it's like we don't exist or something ... they are marginalising the coloured people' (Poachers 2016). In an incisive article outlining the evolution of abalone poaching in South Africa, Johnny Steinberg (2005: 2, 6) notes that the most 'interesting and important' reason for the rise of an illicit market in abalone trade is the 'mutation in the socio-political identities of the coloured fishing communities'. The evolution of a distinctive political consciousness has animated the taking of abalone stock from the water:

> It was a potent combination: on the one hand the expectation that democracy ought to be coupled with the speedy implementation of a just fishing regime; on the other, a deeply held suspicion that the new government would betray the coloured working class. This cocktail of expectations and fears could not have been more propitious for abalone poaching ... Given the politics of the moment, a great many people who had lived their lives on the coastline believed that they were entitled to it, and to a share of the benefits that accrued from harvesting it.
>
> (Steinberg 2005: 6)

The feelings of marginalisation by residents in Hangberg are further manifested in the deep-seated perception that Imizamo Yethu is receiving preferential treatment in terms of governmental and private development projects, while Hangberg is being sidelined. The principal of the public primary school in Hangberg calls it, 'the forgotten community on the hill ... people have forgotten about it. It's sad. And they were the first people in Hout Bay' (Tefre 2010). Indeed one researcher who spent time in Hangberg argues that, 'For the community as a whole, isolation from its neighbouring communities as well as government agencies seems to be the norm' (Tefre 2010: 185).

Marginalisation and autochthony

Taking the stance of marginalisation from the state to the extreme, one group of residents in Hangberg have said they no longer recognise the authority of the state. These residents identify as members of the KhoiSan nation, advocating a nationalism that challenges the foundations of the post-apartheid order through advocating an alternative form of political belonging, based on indigeneity and ontological decolonisation.[5] Hangberg is a site of historical significance for the KhoiSan, where plans for a heritage site have been proposed (Verbuyst 2016). A leader of the KhoiSan movement, talking about Hangberg, explains

> We are the aboriginals of this place ... we are running this community ... whether they say it's illegal ... the ANC is here, the DA is here. Both of them are the cause of the suffering of the community so we don't work at all with them.
>
> (KhoiSan chief 2015)

Their claim to aboriginality is a claim that disrupts the contending rainbow and racial nationalisms that linger in the post-apartheid order. As a KhoiSan activist based in Hangberg explains, the KhoiSan are a first nation meaning, in '"the beginning" ... every person on earth has 15 to 25% Khoi genes in them' (KhoiSan activist 2017). The primacy of the KhoiSan nation is legitimated by a claim of autochthony, and in the spirit of Ethiopianist religious movements and Garveyism, Khoisan nationalism links the claim to be a first nation to that of being the true African people. KhoiSan nationalism thus invokes an idea of the nation that is not just pre-colonial but also decolonial – the challenge is to unthink the categories like 'coloured' and 'black' imposed on African identity.

As would be expected, the KhoiSan have a complicated, ambiguous relationship with the South African state. On the one hand, there is a challenge to the legitimacy of the South African state and nation as colonial impositions on a first nation, but on the other is a desire to secure rights from that state, and thus to be recognised by that state. In terms of the former, activists explain that the current government is 'the same as the white government ... it's an illegitimate government ... no Khoi was present at the drafting of this country's Constitution'. It is 'a foreign regime ... the white man and the black man both in this country are makwerekweres[6] ... the longer our history is kept buried the longer a foreign government can run the land'. Indeed it is 'crazy to give money, rates and taxes from the aboriginal people to this government'. If any actor had legitimate authority it would be the KhoiSan 'king' (KhoiSan activist 2017).

Yet, activists have been engaging with, and hoping to influence, state legislation and policy processes. Thus in February 2016, the KhoiSan revolution party was involved in a protest outside the national parliament in Cape Town, demanding not just land rights but also 'immediate recognition of the Khoisan nation according to the United Nations declaration on indigenous people' as well as 'integration of Khoisan soldiers/Cape Corps into the SA National Defence Force, who were excluded since 21 April, 1994' (Nkalane 2016). Engaging the state has also taken the form of joining workshops and gatherings on KhoiSan culture, or for some, accepting government benefits of land, farms, or business opportunities (including fishing quotas). For those activists we interviewed, however, accepting small concessions from the government was regarded as 'selling out' and denying the larger identity struggle.

While the emergence of KhoiSan nationalism in Hout Bay remains a minority movement in the Hangberg settlement, it constitutes a profound political response to marginalisation. Indeed, its very existence is testimony to the lived experience of exclusion in socio-economic, political, and identity terms that calls into question the right of the Hangberg community to belong in Hout Bay. In this context Khoisan nationalism enables the reinvention of belonging by transforming the lived experience of exclusion as evidence of a distinct and unique social identity: the first nation. This is a social identity that trumps the exclusions of the racialised apartheid and post-apartheid orders through the claim to ethnic autochthony. On this alternative and decolonising logic, rather than being a marginal race, oppressed under apartheid and ignored under democracy, the KhoiSan are a nation that were here first. They are the original inhabitants, and only they truly belong.

Conclusion

Poachers in Hangberg do not have 'a good story to tell'.[7] Yet, they do have an informative and important story to tell. The implementation, or lack thereof, of small-scale fishing policy has led to the increasing criminalisation and marginalisation of an historic fishing community. This community is profoundly disconnected from local participatory and representative democracy, yet subject to poor developmental and strong market governance. The combination of developmental governance through national policy, and market governance, has led to rising informality and marginalisation. Small-scale fishers and their assistants live in Yiftachel's (2009: 88) ever-growing 'grey spaces', trapped in the space between the 'whiteness' of legality (fishing permits) and the 'blackness' of crime and insecurity (poaching). Informality can lead to weakening sovereignty of the state and a concomitant weakening of formal democracy. It can perpetuate a cycle of 'disconnected democracy'.

Living in spaces of informality often leads citizens to respond in two ways: increasing exit from the state or conflict with the state. In our case, poachers have followed both paths. The former is demonstrated by themes such as a well-developed illegal market, the growing move towards Rastafarianism by some Hangberg residents (which is inherently anti-authority) and Khoisan nationalism by others, and the careful avoidance of the law and policy. The latter is seen in the 'Battle for Hangberg' in Chapter 4. It is also clearly expressed as a future option by self-confessed poachers in Hangberg (Poachers 2016) when they say:

> What is the way forward for fishers in Hangberg? Ag nothing ... poaching ... it's all we have ... the last option is to burn all the boats in the harbour, then they will listen. We are not afraid ... there is nothing more that can happen to us.

Notes

1 Abalone, also known as perlemoen (Haliotis midae), is a pale-coloured form of large sea snail, housed in a flat, shiny shell. It is traded live, frozen, tinned, canned or dried.

2 There may be some confusion with this policy as the Department of Trade and Industry has been working with some communities to set up co-operatives, which operate differently from (and are not co-ordinated with) CBLEs (Stern 2013: 14).

3 The EFF (Economic Freedom Fighters) is a left-leaning socialist party advocating for land expropriation without compensation. The Freedom Front Plus is a small conservative political party often associated with the South African white minority.

4 Carriers have an arduous task of moving the heavy abalone (if in their shells) from a drop off point to Hangberg. This can involve climbing over mountains late at night (de Greef 2013).
5 The KhoiSan is the name given to what were several distinct groups of people first living in what later became South Africa. The San were hunter-gatherers and the Khoi or KhoiKhoi were pastoral people. They lived in the southern African region from the late Stone Age, before the southern migration of the Bantu peoples (Diamond 1999).
6 Makwerekwere is an interlinguistic slang word in South Africa for foreigner. It often has negative connotations.
7 This is a phrase the ANC use in election manifestos and campaigning.

References

Allison, E. H. (2001). Big Laws, Small Catches: Global Ocean Governance and the Fisheries Crisis. *Journal of International Development*, *13*(7), 933–950.
Anciano, F. (2014). Non-Racialism and the African National Congress: Views from the Branch. *Journal of Contemporary African Studies*, *32*(1), 35–55.
Chatterjee, P. (2004). *The Politics of the Governed: Reflections on Popular Politics in Most of the World*. New York: Columbia University Press.
Cornwall, A. (2002). Locating Citizen Participation. *IDS bulletin*, *33*(2). https://doi.org/10.1111/j.1759–5436.2002.tb00016.x. Accessed 8 May 2018.
Davis, A., & Ruddle, K. (2012). Massaging the Misery: Recent Approaches to Fisheries Governance and the Betrayal of Small-Scale Fisheries. *Human Organization*, *71*(3), 244–254.
de Greef, K. (2013). Booming illegal abalone fishery in Hangberg: Tough lessons for small-scale fisheries governance in South Africa. Doctoral dissertation, University of Cape Town.
Diamond, J. (1999). *Guns, Germs, and Steel*. New York: W. W. Norton & Company.
Dolley, C. (2006). 'Desperation' sends fisherman on fatal trip. *IOL Online*. www.iol.co.za/news/south-africa/desperation-sends-fisherman-on-fatal-trip-1.271740?ot=inmsa.ArticlePrintPageLayout.ot. Accessed 8 May 2018.
Goga, K. (2014). The illegal abalone trade in the Western Cape. *Institute for Security Studies Papers*, Paper 261, 12.
Hauck, M. (2009). Rethinking small-scale fisheries compliance: from criminal justice to social justice. Doctoral dissertation, University of Cape Town.
Isaacs. L. (2012). Poachers 'have no alternative'. *IOL Online*. www.iol.co.za/capetimes/poachers-have-no-alternative-1281644. Accessed 8 May 2018.
Joseph, N. (2001). Crayfish poachers die in bid to feed family. *IOL Online*. www.iol.co.za/news/south-africa/crayfish-poachers-die-in-bid-to-feed-family-68607. Accessed 8 May 2018.
Kuperan, A. K., & Sutinen, J. G. (1998). Blue Water Crime: Deterrence, Legitimacy, and Compliance in Fisheries. *Law & Society Review*, *32*, 309–338.
Lambrechts, D., & Goga, K. (2016). Money and Marginalisation: The Lost War against Abalone Poaching in South Africa. *Politikon*, *43*(2), 231–249.
Masifundise Development Trust, the Institute for Poverty, Land and Agrarian Studies (PLAAS) at the University of the Western Cape and the Too Big to Ignore (TBTI) network (2012). *Small-Scale Fisheries (SSF) Policy: A Handbook for Fishing*

Communities in South Africa. http://toobigtoignore.net/wp-content/uploads/2014/05/SSFpolicyENG1.pdf. Accessed 8 May 2018.

McDaid, L. (2014). Investigation into the nature of fisher community representation in the development of the small-scale fisheries policy in South Africa, identifying challenges and lessons learnt, and their implications for the perceived legitimacy of the policy. Master's thesis, University of Cape Town.

Mortlock, M. (2017). Update: Hangberg Protesters Torch Construction Vehicle, Intimidate Journos, EWN. http://ewn.co.za/2017/09/14/update-hangberg-protesters-torch-construction-vehicle-intimidate-journos. Accessed 6 January 2017.

Nkalane, M. (2016). Khoisan Leaders Protest Outside Parly. *IOL Online*. www.iol.co.za/news/politics/khoisan-leaders-protest-outside-parly-1981944. Accessed 8 May 2018.

Porter, L. (2011). Informality, the Commons and the Paradoxes for Planning: Concepts and Debates for Informality and Planning. *Planning Theory and Practice*, *12*(1), 115–153.

Schultz, O. J. (2015). Power and democracy: the politics of representation and participation in small-scale fisheries governance on the Cape Peninsula. Doctoral dissertation, University of Cape Town.

SSF (Small-scale Fishers). (2012). Policy for the Small-scale Fisheries in South Africa, Act no 474 of 2012. www.nda.agric.za/docs/Policy/PolicySmallScaleFishe.pdf. Accessed 25 April 2015.

Steinberg, J. (2005). The Illicit Abalone Trade in South Africa. *Institute for Security Studies*. www.files.ethz.ch/isn/99200/105.pdf. Accessed 8 May 2018.

Stern, M. (2013). Sustainable Livelihoods and Marine Resources: How Does South Africa's Policy for the Small-Scale Fisheries Sector Consider Current Challenges on the Ground? Occasional Paper No.166. *SAIIA*. www.files.ethz.ch/isn/176493/saia_sop_166_%20stern_20131231.pdf. Accessed 8 May 2018.

te Lintelo, Dolf. (2016). Enrolling a Goddess in the City. Public Authority and the Mediation of Formal–Informal Relations in Delhi. Paper presented at IPSA World Congress 2016. Posnan, Poland. http://paperroom.ipsa.org/app/webroot/papers/paper_58106.pdf. Accessed 8 May 2018.

Tefre, Ø. S. (2010). Persistent inequalities in providing security for people in South Africa-A comparative study of the capacity of three communities in Hout Bay to influence policing. Master's thesis, University of Bergen.

Verbuyst, R. (2016). Claiming Cape Town: Towards a Symbolic Interpretation of Khoisan Activism and Land Claims. *Anthropology Southern Africa*, *39*(2), 83–96.

Viceland. (2016). Illicit Abalone. *Black Market*, S01E02. www.youtube.com/watch?v=1Wjusc3Oyng. Accessed 8 May 2018.

Yiftachel, O. (2009). 'Theoretical Notes on Grey Cities': The Coming of Urban Apartheid? *Planning Theory*, *8*(1), 88–100.

Interviews

1. KhoiSan activist. (2017). Interviewed by Conrad Meyer. 5 August 2017.
2. Khoisan Chief. (2015). Interviewed by Zikhona Sikota. 6 March 2015.
3. Fisher 1. (2015). Interviewed by Fiona Anciano. 6 March 2015.
4. Fisher 2. (2016). Interviewed by Conrad Meyer. 24 May 2016.

5. Fishers 3 & 4. (2016). Interviewed by Fiona Anciano and Conrad Meyer. 20 May 2016.
6. Fisher 5. (2016). Interviewed by Conrad Meyer. 24 May 2016.
7. PMF leader A. (2015). Interviewed by Fiona Anciano. 17 March 2015.
8. Poachers. (2016). Focus Group by Fiona Anciano and Conrad Meyer. 17 May 2016.
9. Zille, H. (2017). Premier of the Western Cape. Interviewed by Laurence Piper. 26 September 2017.

6 Upgrading Imizamo Yethu

Contests of governance and belonging

On 11 and 12 March 2017, a massive fire swept through the informal section on the upper slopes of Imizamo Yethu, commonly known as Dontse Yakhe, destroying 2,194 structures and displacing an estimated 9,700 people (CCT 2017a, 2017b). The fire decimated nearly one-third of the area of Imizamo Yethu, and razed to the ground all the informal structures in the most densely populated section. In the immediate aftermath of the disaster the City, supported by NGOs like the Red Cross, distributed emergency relief supplies and erected marquees on the sports fields opposite the entrance of Imizamo Yethu (see Figure 6.1). These efforts were greatly aided by substantial donations and others forms of material and logistical support by the wealthier residents of Hout Bay.

The fire was the single greatest disaster ever in Hout Bay, and the largest in a series of fires in Imizamo Yethu's history. The worst fire previous to this was in 2004 and destroyed an estimated 570 homes leaving 2,500 people homeless (Macgregor et al. 2005). To add insult to injury in 2017, a second fire swept through Imizamo Yethu in April, destroying a further 112 structures and displacing 500 people (CCT 2017b). In response to the disaster the City initiated a process of upgrading called 'super-blocking'. 'Re-blocking' informal settlements means rebuilding the settlement in 6m by 6m plots, with freestanding structures made of fire-retardant materials and with a gap between them to reduce the chance of fire spreading (City official 2 2015). 'Super-blocking' means also building a road, and providing proper amenities for water, electricity, and sanitation.

In developing the plan to 'super block' Dontse Yakhe, the City set up an elected working committee from a 19-organisation stakeholder forum in Imizamo Yethu that met regularly in the weeks and months following the fire (CCT 2017b; City officials 2017). Yet, despite this process of consultation, opposition to the super-blocking grew among some residents of Dontse Yakhe, leading to the burning of the ANC office, as well as the houses of the Community Development Worker, and two ANC leaders in July 2017 (Cronje 2017; de Villiers 2017; Community leader 2 2017). Then, on the 15 August 2017, around 2,000 protesters marched to City Hall, where a memorandum was handed over to a representative of the Mayor (Adriaanse 2017).

Figure 6.1 Aerial photograph of Dontse Yakhe after the fire

Frustrated at the mean-spirited reply by the Mayor, a week later the Dontse Yakhe protesters closed the road to Wynberg by chopping down trees into the road and burning tyres (Mediator 2 2017).

Superficially, this protest against super-blocking makes little sense, and was publicly blamed on 'misinformation' and 'meddling' by leaders of Dontse Yakhe (de Villiers 2017). Others point to the heavy-handed and dismissive approach of the City, which has failed to consult directly with the residents of Dontse Yakhe (Mediator 2 2017). Although in part attributable to these factors, the resistance to super-blocking is also an attempt to defend a range of informal practices around property and livelihoods threatened by the mono-lithic approach of developmental governance that sees upgrading as only about shelter. Indeed, it is through the frame of contending forms of gov-ernance that the tensions that grew between various leaders within Imizamo Yethu after the fire makes the most sense.

In what follows, the chapter traces the politics around settlement upgrading in Imizamo Yethu through two sets of lenses. The first, as indicated above, is the lens of developmental and informal governance, demonstrating the diver-gent logics of these forms, and the different ways that residents engage with governance. The second, is the different meanings of being and belonging in Hout Bay expressed by contending groups in contests over settlement upgrading in Imizamo Yethu. These notions of belonging, framed in terms of race, party, and nationality, are illustrated through three contests linked to upgrading: the 'battle of the green belt', the protest against the Disa School, and the resistance to super-blocking in Dontse Yakhe.

While there is no necessary relationship in the tensions between these two sets of dynamics (developmental governance versus formality; and three discourses of belonging in Hout Bay), in the case of Imizamo Yethu they intersect in ways that unsettle the local order and disconnect it from formal democracy. Thus, developmental governance in Imizamo Yethu tends towards party capture, and therefore Chatterjee's (2004) 'political-society' forms of patronage, nepotism, and even corruption. Informal practices are closer to Bayat's (2000) account of quiet encroachment. Neither approach sits comfortably with formal democracy in general, but in the case of Imizamo Yethu these tensions overlay conflicts of race relations between Imizamo Yethu and the Valley, and between residents in formal housing and those in informal housing. The net effect is a disordering of local rule that has a further effect of deepening the legitimacy crisis of local leaders (manifest in the factionalism in SANCO), and weakening the capacity of Imizamo Yethu to respond to crises like the 2017 fire with a united voice.

Facilities, race, and a suburban future

This chapter opens with a moment of protest against informal settlement upgrading after the 2017 fire, but there is another instance just as important for understanding governance in Imizamo Yethu, and that is the protest in November 2011 against the opening of the Disa Primary School. Starting near the entrance of Imizamo Yethu, a smallish crowd of around 100 people marched through the township, stopping at the large, double-storey brick house of the 2007 SANCO leader,[1] a public supporter of the school, before continuing to the school grounds. According to the protesters, the 2007 SANCO leader 'had not consulted' about the school, and they claimed the community needed housing rather than a school. In reality then, the protest was not really about the school, but rather about different visions for the upgrading of Imizamo Yethu, and even more profoundly about who really belongs in Hout Bay, and on what terms.

When Imizamo Yethu was founded in 1991 in the dying days of apartheid rule, 18 hectares of land were put aside for housing and 16 hectares for 'community facilities and a green belt' (Harte, Hastings, & Childs 2006). Some civic organisations like Sinethemba, that claimed to represent the 'original' multi-racial group of families that settled in Imizamo Yethu, argued against the growth and densification of the settlement and wished to see 16 hectares used for 'its original purpose' of community facilities. Sinethemba was supported vociferously in this claim by the middle-class and white-dominated Hout Bay Ratepayers Association, who also wanted to contain the growth of Imizamo Yethu, as Hout Bay was 'too full' already, and more poor residents was associated with higher crime rates, disease, and environmental pollution.

Against this view, from the mid-1990s the South African National Civic Organisation (SANCO) lobbied for the use of this land to build more houses, partly due to the clear shortage of such, given the high rate of immigration, but also in the name of racial integration. SANCO argued that building specific facilities for Imizamo Yethu would consolidate, rather than challenge, the racial segregation of Hout Bay.

Built between 2009 and 2011, the Disa Primary School admitted its first intake of students in 2012. The school has excellent facilities, which were paid for by German benefactors who live in Hout Bay, Dr Andreas and Susan Streungmann, owners of Hout Bay Manor Hotel. Indeed, these benefactors help pay for running costs on an annual basis, as managed by a Board on which they sit, with the result that poor children in Imizamo Yethu and Hangberg can afford to attend. The Disa School clearly provides an opportunity for at least some residents to secure an excellent education for their children at very affordable rates. The Disa Primary School was officially opened by Western Cape Premier Helen Zille in November 2012, and the Hout Bay Ratepayers and Residents Association (HBRRA) attended the event. Reporting in the monthly newsletter, 'Hout and About', long-standing HBRRA chair Len Swimmer (2012c) stated:

> On leaving the opening of the Disa Primary School, we were suffocated by the stench of grey water, most probably unadulterated pure raw sewerage running down through the alleys next to the shacks. We had just left a 1st World School, in a 3rd World squalid environment. Our Association pledges to get change here and to do everything we can to force the Authorities to transform this terrible situation without further delay; to make this area into a green leafy suburb of which all of Hout Bay can be proud. For this fight we need the community's financial help and support, to take this Administration again to the High Court if need be, as it is a crime against humanity and contrary to the South African Constitution to witness the squalid conditions people have to live in on this mountainside.

This statement captures the essence of the Ratepayers' views on Imizamo Yethu. In the preceding 10 years, the HBRRA had tried everything in its power to limit immigration into Imizamo Yethu, and to advocate for housing development that yielded a 'green leafy suburb'. There were several elements to this vision. First, was ensuring 'decent' sized formal housing. Using South Africa's standard low-cost housing model of 40 units per hectare, this would mean around 720 houses on the 18 hectares originally designated for houses. Second, was the provision of facilities like a number of schools and a green 'buffer zone' on the remaining 16 hectares. Third, was the removal of all informal housing, especially on the slopes above the formal settlement that later became Dontse Yakhe.

Central to the politics of lobbying for this vision of Imizamo Yethu was the pressuring of the City of Cape Town, through formal channels, to enforce the original 1991 plan for Imizamo Yethu, and then, when unsuccessful, turning to the courts to rule on land use, zoning, and environmental objections. As the struggle over the development of Imizamo Yethu intensified in the mid-2000s, the HBRRA turned especially to concerns around the lack of sanitation and its impact on health and the environment, issues reflected in the quote above, as the main concerns to justify its actions. Somewhat ironically this conservative 'politics of sewerage' prefigured the insurgent 'poo flinging' of subsequent radical social movements like the ANC-aligned *Ses'khona people's movement* in Khayelitsha of the 2010s.

Initially the HBRRA seemed to be successful in having its case heard by the City of Cape Town. Thus, by 2002 the executive committee of the then ANC-run city agreed that overcrowding in Imizamo Yethu would be solved by removing people to available places in Mitchells' Plain and Blue Downs (Zille 2007). The City even proceeded to get a court order to begin moving newcomers out. In the words of Len Swimmer (2012a), in 'the following month trucks with low-bed trailers lined up in Hout Bay Main Road to move settlers to a Temporary Relocation Area (TRA)'. However, on 4 June 2003, City Mayor Nomaindia Mfeketo declared a one-month moratorium on shack demolitions in Hout Bay. Mfeketo's city manager, Wallace Mgoqi, stated that they would instead seek 'a more humane solution to the overcrowding' (Zille 2007).

In the event, nothing occurred until February 2004 when a major fire swept through the informal settlement, leaving an estimated 570 shacks destroyed (Monaco 2008). Attempts were made by Niall Mellon to build 100 new houses in the places gutted by the fire, but were delayed until the end of 2005. In the interim, those displaced by the fire were left in tents and community halls, and eventually began to build informal structures above the formal boundary of Imizamo Yethu in Dontse Yakhe (Figure 6.2). When the city started removing trees from the 16 hectares to resettle the families, the Hout Bay Ratepayers Association and Sinethemba obtained a High Court interdict to stop the building.

In response, the Provincial MEC for Housing, Richard Dyantyi, on request from the City of Cape Town, declared that the remaining 16 hectares of land would be rezoned for housing, and amended the Less Formal Township Establishment Act (n113/1991) to overturn the High Court ruling. In 2006, following a study of water quality in the Disa River that showed dangerously high concentrations of E.coli, the Sinethemba Civic Association and the Hout Bay and Llandudno Community Policing Forum took further court action against the City to remove the informal settlers from Dontse Yakhe.

Following the electoral success of the Democratic Alliance in 2006, Helen Zille became Mayor of Cape Town and paid for a facilitation and consensus process with the various leaders of Imizamo Yethu and other communities in Hout Bay. This process was facilitated by the Institute of Justice and Reconciliation, and established the 'IJR Principles' by mid-2007. These

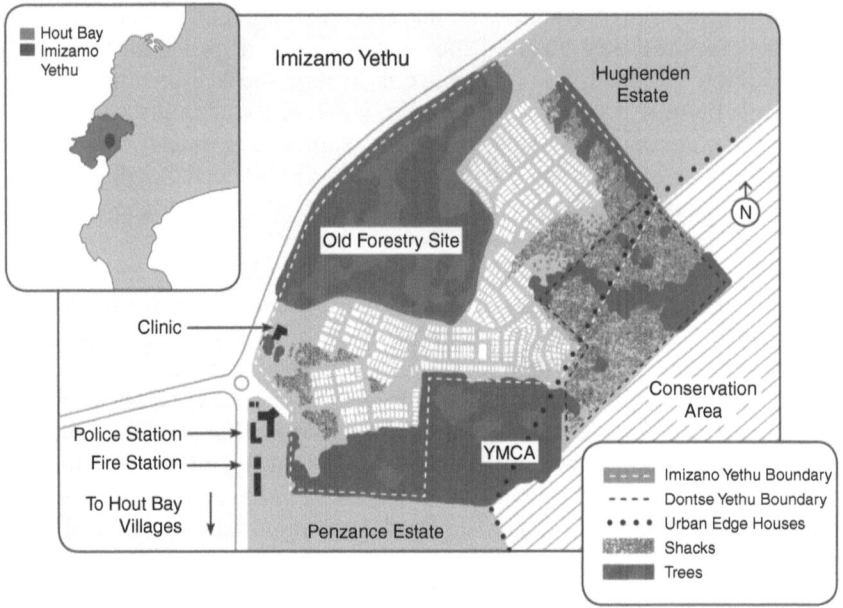

Figure 6.2 Map of Imizamo Yethu 1991 development plan (based on Harte et al. 2006)

included, in the words of Len Swimmer (2012b), agreement that 'movement of some residents is inevitable', that 'voluntary movement of people is also an option for consideration' and that 'no person will be moved unless acceptable housing conditions are provided in another setting'. Informed by these principles, in 2008 the City proposed four development options to the Hout Bay community with different combinations of housing and community facilities. Having received comment from the community, the City presented a fifth option ('option 5') in late 2008. This provided for 46 single residential units, and 1,000 apartments (of 40m² each). It included a primary school but excluded a high school. Province authorised these plans in November 2009.

Objecting to these plans, the Hout Bay and Llandudno Environment Conservation Group (ECG), closely allied to the HBRRA, attempted unsuccessfully to have the plans set aside by the High Court. Notably, among the objections were the failure to plan for a high school as well as a primary school, and the failure to retain 'the belt of trees along Main Road, as indeed required by the applicable environmental authorisation' (Hout Bay & Llandudno ECG v Min Local Government 2010: 6). In this way, the desire for specific facilities for Imizamo Yethu and a visual buffer were reiterated. Further, the HBRRA were somewhat upbeat about their influence on the new plans. Thus Swimmer (2012b) notes:

Having spent R330,000 in legal costs we now have at least a hope that this IY development will be what the residents want and have a human feel, whereas the layout proposed by the City's planners would have been a sterile wasteland of cheek by jowl dwellings and would have resulted in the destruction of the broad band of trees along Main Road to make way for a wrongly located 'service road' parallel to and just metres away from Main Road.

In respect of the informal settlement of Dontse Yakhe, however, little changed. In mid-2011, the City made application to the High Court to evict all those occupants who resided within 15 metres of the water mains that run down its northern edge. The matter was heard on 24 October 2011 and the Court ordered that the parties engage meaningfully about whether alternative accommodation could reasonably be made available to the affected persons. Negotiations broke down, and the City identified the affected dwellings on its own, although no relocations took place. A key reason for this is, apparently, the lack of alternative space for a temporary relocation area in the City.

Critically, at the heart of the HBRRA's vision of Imizamo Yethu as a 'green, leafy suburb' evident in the pages of 'Hout and About', is a clear rejection of any form of informality, whether street trading, street mechanics, shebeening or the litter, clutter, and noise that currently marks life in the settlement. However, these views are not just about the environment and conservation, but also about class and social values. Indeed, a key assumption of the HBRRA is that the problems of Imizamo Yethu would disappear if immigrants were simply wealthier. For example, in a letter to the *Cape Times* on 5 October 2009, Eric Shaug, an HBRRA executive committee member, argued that the technical difficulties of building on slopes means that 'only the rich can afford to build on sloping sites. One might argue that the poor should not be discriminated against, but there is a finite amount of money available for their housing.'

Further, writing in 'Hout and About' in February 2010, Su Ball argued that the City's development plan for Imizamo Yethu would create a 'visual blight ... scarring the slopes of Skoorsteenkop. Without sufficient job opportunities and educational facilities, people will live squashed together without hope for a better life, a breeding ground for sickness and crime'. She added:

Such a plan for high density walk up flats robbed of screening trees and a gradual interface with the surrounding suburbs will create a visual blight in the centre of the Valley. Instead of tourists being attracted to the area they will be repelled. Tourist numbers will dwindle causing businesses dependent on the tourist industry to collapse having an economic domino effect on Hout Bay of joblessness, crime and squalor in an ongoing downward spiral. Wealthy landowners and businessmen will abandon Hout Bay diminishing the rates base of the City. Let Hout Bay fulfil its destiny not as a slum but as a thriving tourist village and an asset to the broader Cape Town economy.

For the HBRRA, Imizamo Yethu ought to become a sprawling, middle-class suburb, not a poor high-rise settlement. Notably, the HBRRA's turn to the environment as a key reason to intervene in Imizamo Yethu, and to resist immigration by poor, black migrants into Hout Bay more generally, mirrored its conservation discourse around the Table Mountain Nature Reserve in resisting the imposition of the Chapman's Peak Toll Road by private developers, described in Chapter 8. Conservation, then, is invoked by the champions of the Valley against both private developers, informal practice, and developmental governance in Hout Bay. Indeed, by 2013 the HBRRA's official slogan was 'The environment is, and will always be, our priority'.

Housing, the ANC, and factionalism

In contrast to the HBRRA, SANCO in Imizamo Yethu has long held a very different vision of appropriate development and the place of poor, black people in Hout Bay. At the heart of the vision is first, the use of all 34 hectares of land in Imizamo Yethu for housing (see Figure 6.3); second, the concomitant integration of the poor, black residents of Imizamo Yethu into the schools, clinics, shopping areas, and other public facilities of Hout Bay. Third, there is the view often articulated by leaders for the recognition of all as 'Hout Baynians', and inclusion in the settlement, rather than treating Imizamo Yethu as a reserve of poor, black labour segregated off from 'white' Hout Bay (Community leader 2 2012, 2017). The call for inclusion should not

Figure 6.3 Aerial photograph of housing types in Imizamo Yethu

be confused with a rejection of racial ways of thinking, however; indeed, race is the primary means used by all leaders to frame local politics.

These claims are complicated by the fact of the existence of rival civic organisation Sinethemba to represent the residents of Imizamo Yethu. However, even if Sinethemba could claim to be the first civic in Imizamo Yethu, it is clear that with the influx of a much larger group of, mostly isiXhosa-speaking, people in the 1990s, many of whom were migrants from particular areas of the Eastern Cape, SANCO soon had more support. In addition, as Monaco (2008) notes, from its emergence in 1991 the Imizamo Yethu Civic Association, which affiliated to SANCO in the mid-1990s, lobbied for the 16 hectares of land reserved for community facilities to be used for housing. Thus, virtually from the outset, divergent visions of the development of Imizamo Yethu were at play.

Further, as Sikota (2015) demonstrates, SANCO played a key leadership role in bringing housing to Imizamo Yethu from the late 1990s. Thus, in 1997, SANCO attempted to access the new government housing subsidy to build structures on the assigned plots under what became known as the Makukhanye housing project. Assisted by an NGO, the Development Action Group (DAG), SANCO applied for government funding under the People's Housing Process (PHP) established in 1998, later amended into the Enhanced People's Housing Process (EPHP) in the National Housing Code of 2009.

In terms of the PHP, DAG's role was to assist the community in forming a project committee, develop a plan, and facilitate the engagement with the state. According to key members of the Makukhanye project committee, the constitution of the committee, the development of the plan, and initial engagement with the state went well (Sikota 2015). While Sinethemba members dispute how welcome they were made to feel in this process, it remains clear that the SANCO leaders did enjoy more support and that these formal processes were followed closely by the City and DAG. While the project application was successful, it began to run into implementation problems before long. A key issue was the inability of beneficiaries to comply with the 'sweat equity'[2] requirements of a PHP. A Makukhanye leader explains:

> because most of the people are working during the day, they cannot be building houses and sort of working together as Ilima [a co-op] to build the houses, so we decided to have a contractor, [and move] from People's Housing Project to a managed PHP because it's managed if there is a contractor.
>
> (Sikota 2015: 64)

In light of these constraints, the Makukhanye housing project transitioned to a managed PHP in 2002. Initially, the international NGO, Habitat for Humanity, was the contractor, not only building the houses but also providing loans to some beneficiaries for building their houses. However, as the project unfolded, the building process was slow and the government subsidy was insufficient to build the kind of houses that the beneficiaries wanted, even with Habitat's help (Sikota 2015). A major breakthrough for the project was the introduction of Niall Mellon as a possible contractor in 2002. An Irish property developer who came to Cape Town as a tourist in 2000, Niall Mellon was inspired to help with the housing project by his first-hand experience of Imizamo Yethu.

Mellon's intervention in the Makukhanye housing project transformed it. He set up the Niall Mellon Township Trust (NMTT), which took over the implementation of the project from the DAG and Habitat for Humanity. One of the consultants who compiled the NMTT assessment report believes that 'Niall Mellon just wanted to build, he wanted to see progress and that is why he started working on the project before the government subsidy was given' (Sikota 2015: 69). The project ran from 2003 to 2005 during which time about 448 houses were built with the involvement of the beneficiaries, Niall Mellon, the state and other stakeholders.

Central to the ability to speed up the build was Mellon's capacity to bridge finance the building while enduring the typically long wait for government subsidies to come through. The project continued Habitat's approach in which half of the cost of the new homes was paid for by the NMTT, and the other half financed through a combination of state subsidy and beneficiary payments (Rangasami & Gird 2007). However, given the delay in securing government subsidies, NMTT offered an interest free loan to beneficiaries to cover their costs, and reclaimed the subsidy directly from government. As a community leader noted:

> Mellon was the preferred candidate because he have all the resources. … the subsidy at that time was R13,900 so we started rolling on … the advantage of Mellon was that he was going also to put some money in the project as an addition on the condition that the community will pay back … and the specification, which means the size of the house at that time was 36m² and then Habitat was coming with 40m² house but Mellon was also prepared to work beyond those because … the smallest house from Mellon its 38m² and I think it's maybe five houses but the rest its 50m² house. We've got 68, we got 72, we got loft.
>
> (Sikota 2015: 82)

Thus, in addition to bridge financing, Mellon also allowed recipients greater architectural choice in terms of house design, a view confirmed by a focus group of recipients (Sikota 2015). The NMTT also assisted with the building of houses by organising Irish volunteers to work alongside locals in housing 'blitzes' every year. Notably, though, it was around the issue of the recipients of housing that conflict began to manifest. Thus, according to one Sinethemba member:

> I went to some of the meeting but they did not want us as Sinethemba to be part of the project. They wanted to side line us … first of all it was the

things we put forward in the meeting, especially things that we were not happy with regarding the leaders of the project. Things we were trying to prevent. For example, things like new residents getting houses on other people's sites just because ... example, Maduna left here and he had a house but we don't know where he went and another guy who also came from Princess Bush left here and when he came back, his house was given to someone else.

(Sikota 2015: 74–75)

The selection of recipients was not a process directly managed by the NMTT but was run through the SANCO-aligned PHP leaders, in association with city officials. It seems the PHP project leaders had significant influence in the process. Hence, when the NMTT suggested they hire a project manager, the civic leadership identified a local resident for this post. They also secured from the city office a space for a Niall Mellon housing office in Imizamo Yethu. This provided beneficiaries with an opportunity to get information on the project, access to project staff, and a place to make their interest free loan repayments.

While the issue of who made the recipient list, and the rivalry between SANCO and Sinethemba, did manifest in public debate as a form of nepotism, including the NMTT 2003–2005 project assessment report (Rangasami & Gird 2007), it does seem that the first phase of the Mellon project was a clear success. No less than 448 houses were built in 3 years, ranging in size from 48m² to 72m², with clearly positive implications for the health, education, safety, and especially a sense of belonging of recipients. According to one: 'I feel like I'm in heaven. I have found my heaven on earth with my new house.' Another added, 'Niall Mellon is like Moses from the bible who released the Israelites from persecution in Egypt and took them to Canaan, the land of milk and honey' (Rangasami & Gird 2007: 8). Finally, and critically, SANCO leaders played a key role in initiating and sustaining engagement with the state and NGOs to get the project done.

Within a few years of the highpoint of the Niall Mellon housing project of 2005, however, conflict about beneficiaries and leadership of housing projects reached new lows. In 2007, a new set of SANCO leaders took office in Imizamo Yethu. This leadership group was quite small and limited to the chairperson and two or three closer associates. While this leadership was very effective in participating in the structures of local governance in Hout Bay, such as the ward forum, the Hout Bay partnership, and the like, it soon found itself in conflict with others in the larger ANC network in Imizamo Yethu for embracing the HBRRA's development vision. In particular, the ANC opposed the building of community-specific facilities on the 16 hectares that SANCO had long lobbied for as space for housing. These issues bubbled under for a few years around the state-led Masakhane housing project before coming to a head with the building of the new Disa primary school in 2011.

The Masakhane Bantu PHP housing project was a state-led sequel to the Niall Mellon project, designed to upgrade Dontse Yakhe by providing formal housing for 143 families. Conflict emerged between the PHP committee, aligned to the ANC, the City of Cape Town's housing department, and the new SANCO leadership, about who should be on the beneficiary list. Thus, in 2009, when the keys to about 68 of the houses were handed over to recipients, at least 20 families found that their houses were already occupied. These occupants were families on the original housing list compiled by the project committee who were then removed by the City based on objections that they did not qualify for the PHP programme. The City then gave title deeds to families who had lived in Imizamo Yethu for longer, but the occupants resisted on the basis that the PHP had given them the land and they had used their subsidy to build the houses (Mukadam 2013). At the centre of this conflict were allegations about corruption in the PHP around the allocation of places on the housing list, and the 2007 leadership of SANCO took the side of the City against the PHP committee. The 2007 SANCO leader explained:

> These people were … illegally accommodated. They are not … long[-standing] … citizens of Imizamo Yethu … they bribe the people in order for them to be accommodated. I have been saying this word bribe time and time again because I know exactly what I am talking about, I have been engaged with the city, I have been engaged with the province. They [the PHP] decided to tell me that the city has made a mistake but I said to them, call the meeting so that the city and the province must have to explain this to us. They never call that meeting, why they didn't? It's because they know that they continue with this corruption from the Niall Mellon houses but nobody follow them on the corruption that was taking place so they decided to continue with it and then we said no its enough.
>
> (Sikota 2015: 91)

Not surprisingly, the ANC-aligned leaders of the 2011 SANCO and their allies deny these allegations, and indeed return them, pointing to the fact that the leader of the 2007 SANCO works as a gardener in the Valley and yet has a large double-storey house in Imizamo Yethu (ANC Youth League leader 2012). Where the 2007 leader points to the generosity of his employer, his rivals point towards corruption, especially from wealthy interests in the Valley. Like Sinethemba, the 2007 leader of SANCO stands accused of being a sell-out to white and DA interests. Notably, over years of fieldwork we had conversations with a wide range of residents who shared a perception that all the SANCO factions were in some way corrupt in their handling of housing.

The SANCO leadership conflict engendered by the Disa School decision continues today. While there have been at least three further SANCO leadership elections since 2011, the 2007 leadership refused to recognise the legitimacy of the subsequent iterations of leaders, appealing to the district and provincial levels of SANCO which, it turns out, are equally factionalised and thus have been unable to resolve the dispute. Thus, as the 2007 leader states, 'I am the chairperson of SANCO. The term is taking 2 years but because of problems from the national and provincial, we decided not to continue with the election now until we sort out the problems.' He adds:

> the reason why they keep saying, there is this word, we are not legitimate SANCO is because they want to replace me because I don't allow any wrongdoing to happen. Because they want some short cut, they want some people who are going to agree with them and allow for the corruption to go on. How long is the corruption going to go, how long is (sic) our people going to suffer because of selfish people who want to lead this community?
>
> (Sikota 2015: 90)

Against this view, the current ANC leadership sees the 2007 SANCO as a front for the interests of the Valley, an entity that has been bought, and is used to legitimate white and DA interests in Hout Bay. Hence:

> I was a witness yesterday, I was here when we bless SANCO ... There was no old SANCO, the one of [2007] ... was defeated long time, six years ago. He came and he was out voted in this hall with the region and province of SANCO being present [in 2013]. He was out voted after an election and stood up thank and blessed the new leadership ... after that he contested against the same leadership. He is an independent SANCO leader with no mandate, no election authorities. He is being used by the white people, he doesn't want ... he want to use the SANCO name in order to be seen as with the poor ... for him to join the Hout Bay Ratepayers, which is white. So the whole strategy used by those white [laughs] is to use this ... black associated name, SANCO in order to deal the dealings that are anti the people of Imizamo Yethu, that are anti those blacks so that they can have a big say in convincing the Mayor, that SANCO, Hout Bay Ratepayers, all these civic organisation have agree in one thing.
>
> (Sikota 2015: 89)

It is very clear that the ANC in Imizamo Yethu, and the dominant faction of SANCO aligned to it, perceive the actions by the HBRRA as a racist attempt to control the number of immigrants into Hout Bay to 'keep it white', and also to segregate black from white by creating blacks-only facilities and a visual buffer of trees between Imizamo Yethu and the Valley. Indeed, in a conversation with an ANC leader in 2015, we asked whether the large number of foreign nationals in the Valley diluted the alleged racism of white South Africans. He snorted with derision. 'They're even worse, especially the Germans' (ANC leader 2015). Further, the collaboration between Sinethemba and the Ratepayers has led SANCO to see Sinethemba as a stooge of the Valley and its allegedly racist agenda. As one SANCO leader put it, 'Sinethemba was born through the intervention of the outside Hout Bay, the white people in particular who buy way to develop the 16 hectares for the people of Imizamo Yethu' (Sikota 2015: 86).

Two insights are offered by this history of contests over formal housing in Imizamo Yethu. The first is the fact that although the formal democratic system does not establish political representation at any level below the ward, all forms of developmental governance require representation at the level of the settlement. Thus, whether through the PHP, the Masakhane Project, or the current super-blocking process, settlement upgrading requires community engagement with representatives sitting on steering committees, development trusts, consultative forums, and the like. This creates opportunities for gatekeeping, brokerage, or mediation linked to the distribution of resources in a poor area that can amount to a form of career in poor settlements (von Lieres & Piper 2014).

The second insight is that the local leaders who sit on these structures are not formally elected nor selected in legally enforceable ways analogous to the election of a ward councillor. Local leadership positions are thus always vulnerable to contestation, and the authority of local leaders often depends on popularity unless they are able to mobilise some coercive capacity of their own. In the case of Imizamo Yethu, as in the rest of South Africa, the legitimacy of community leaders is shored up by invoking the association with the ANC as the liberator of the black oppressed in South Africa. While this notion of 'party-society' does usually restrict leadership of poor, black settlements to ANC-aligned leaders (Piper 2015; Piper & Anciano 2015), as the case of factionalism in SANCO shows, it does not prevent rivalry for office emerging from within the ANC itself. In Imizamo Yethu, the racialised conflict over the development of the 'green belt' in Imizamo Yethu has fed the factionalism that currently weakens community leadership.

Considered together, these two factors of (i) gatekeeping opportunities created by developmental governance, and (ii) the party capture of representational claims, do approach Chatterjee's (2004) characterisation of 'political society' in the Global South. That is, rather than treating people as democratic individuals bearing rights, residents of poor communities are managed as populations requiring development, and thus exist in patronage relations with the state, mediated by political parties. In this regard, it is notable that SANCO's self-conception as a mediator between society and party-state explicitly fits this description (Zuern 2011). The party capture of representation in developmental governance and tendencies to patronage politics has parlous implications for community democracy by limiting inclusion on a partisan basis. As noted in Chapter 3, no settlement upgrading policy allows for a community-level definition of the substance or standards of upgrading – a point made at the beginning of this chapter – and democratic elements are restricted to aspects of project implementation. What the argument earlier in this chapter shows is how these opportunities to influence implementation are often captured by partisan actors who tend to pursue their own ends rather than more collective ones. This granted, what the party wants is almost never entirely what they get.

Informality, livelihoods, and nationality

As noted in Chapter 3, access to housing in Hout Bay is framed differently depending on the forms of governance that construct the settlement. Thus, for wealthy residents, access to housing is about the individual purchase of property through market governance, whereas for poor residents access to housing is through developmental governance that provides shelter for a needy population. Had urbanisation stopped in 1991, all residents of Hout Bay would be in either private or state housing today. However, as noted in Chapter 1, with a rate of urbanisation of 3 per cent per annum into the City of Cape Town and of 10 per cent per annum into Hout Bay, around 1,000 new poor, black migrants per annum have settled in Imizamo Yethu. Excluded financially from the market, and undersupplied by developmental governance, poor migrants have built or rented informal structures that constitute more than 75 per cent of all housing in Imizamo Yethu.

This reality gives Imizamo Yethu its 'grey' character between formal and informal, legal and illegal (Yiftachel 2009), and this greyness extends beyond housing to services. Until the fire of 2017, most electricity connections were illegal, sanitation took place outdoors, and there was no road through to the top of the settlement. Furthermore, as illustrated in Chapter 3, informality extends to livelihoods too, from residents who trade on the street in Hout Bay, to informal micro-enterprises spread throughout the informal settlement at a ratio of 1 to every 10 households. Notably the most common of these is the shebeen, of which there were 182 in Imizamo Yethu in 2013 and of which only a handful have a liquor licence (Figure 3.4). Furthermore, and this is important for understanding the resistance by some to super-blocking, while most proprietors of shebeens are South African, a significant number are foreign nationals. In Dontse Yakhe, in particular, Ovambo residents from Namibia and Angola are known to run a few large shebeens located on an interior courtyard surrounded on all sides by shacks so that they remain concealed from public view (Community leader 2 2017).

Thus, to return to the protest against super-blocking: in addition to the allegations of poorly managed consultations with residents of Dontse Yakhe (Mediator 2 2017), residents have legitimate concerns about the costs to their livelihoods of formalisation, not least as the city approaches the issue exclusively in terms of providing shelter for the needy. The plan for Dontse Yakhe includes building structures on 6m x 6m plots, providing services like electricity, water, and sanitation, and building a new road. But, there is no place in the plans for the shebeens, haircare, food retail, and other current livelihood uses of informal structures (see Figure 6.4). For example, contrary to the common practice of building an extra room for a shebeen or house shop, City official 2 (2015) from Informal Settlement Upgrading stated 'We cannot

Figure 6.4 Informal life in Dontske Yakhe

allow it because it could densify our informal settlements'. Thus, for those groups who run shebeens, especially foreign nationals who may not enjoy full rights under law in South Africa, the spatial design of super-blocking will likely threaten a livelihood.

In addition to livelihoods practices, another constituency potentially threatened by super-blocking is landlords who own multiple shacks, or rent out rooms in large, sprawling shacks. Indeed, according to a community leader (2017) in support of super-blocking, the main leader of the protest is someone who has a 10-room shack that she stands to lose should the re-blocking proceed. With each room renting at R500/month minimum, that is a monthly income of R5,000 (excluding electricity charges) under threat. While no research exists on the nature of ownership in Dontske Yakhe, residents spoke of the existence of landlords who own multiple shacks or large shacks with multiple rooms. This emergent class of informal entrepreneur stands to lose at least some income from investing in property informally. Lastly, the City's plans for Dontse Yakhe include building a new road that, per regulation, must be at least 8 metres wide. Most perceive this as taking significant space in a crowded settlement, and therefore implying the relocation of at least some current residents, despite the assurances of the City (Mediator 2 2017).

In short, the framing of super-blocking as about providing shelter for a needy population ignores the multiple potential meanings of a 'shack' under conditions of informality. The actual structure itself is but one resource that an informal shelter can provide. A further meaning of a shack is access to land that, albeit informally, has a value over and above the structure itself. Thus, in Imizamo Yethu roughly two-thirds of the purchase price of the cheapest shack (R15,000) is the land, and one-third (R5,000) is the cost of the structure. The distinction between land and shelter is important, as land can be rented for use that is not shelter. This brings us to the third potential meaning of a shack, and that is use of urban land for livelihoods, whether an educare centre, hair salon, spaza, house shop, or shebeen. Finally, if the shack has a number, the owner is on a housing list with the City of Cape Town, and has a right to a formal house at some point in the future. In short, there is far more to an informal structure than just shelter, and framing development primarily in these terms is potentially a threat to land, livelihoods and a place for the poor to belong.

Informality also matters in terms of belonging because a significant proportion of informal settlers in Imizamo Yethu are not South African. Migrants from the rest of Africa have been part of Imizamo Yethu nearly from the start, with a significant Ovambo-speaking population from Namibian and Angola working in the fishing industry. This has caused tensions with black South Africans. According to a number of respondents, there was an attempt to expel Ovambo migrants from Imizamo Yethu in 1992 that was unsuccessful, as the migrants armed themselves and defended the attack vigorously (Refugee 1; Refugee 2; Community leader 2). Some, including officers we interviewed at the Hout Bay Police Station, suggest that a number of these migrants were formerly part of koevoet or 32 battalion of the apartheid-era South African Defence Force. Whatever the truth of these perceptions, the vigorous defence of the Ovambo population in 1992 garnered them a fearsome reputation that meant that they were not targeted in the xenophobic attacks of 2008.

There is no doubt that the high point of xenophobic conflict in Imizamo Yethu took place in 2008 when, following the wave of attacks that rippled across the country, a similar pogrom took place in Hout Bay. According to respondents living in the settlement at the time, the initial response of the community to the attacks elsewhere in the country was muted. However, as the wave broke closer to Hout Bay, washing through the iconic Cape Town township of Khayelitsha, dynamics began to shift. Hence, a local Zimbabwean leader reports (Refugee 1 2012):

all of a sudden I was no longer welcome in SANCO meetings. This was strange as I used to attend all of them, and I had good relations with all the leaders. I remember at the start of the week I attended a meeting where we discussed how to stop xenophobia, but then later in the week when I went to the follow-up meeting they told me to go away. The next day the attacks happened.

By the standards of the rest of the country, the 2008 xenophobic mobilisation in Imizamo Yethu was mild. Early on the morning of Saturday 14 June 2008, local leaders, including some SANCO people and especially street committee members, circulated the word that all foreigners were to leave Imizamo Yethu. As one respondents recalls: 'Panic set in. We had seen what had happened in other places, even in Cape Town'. By that evening almost all foreign residents had fled Imizamo Yethu. However, within 48 hours all were back. 'We heard that once the shops closed and people have to pay to go to checkers to get bread that they wanted the Somalis back ... they got the owners to call them back' (Refugee 2 2012). No one was killed in this incident, and we witnessed vigorous debate over whether any actual violence took place. According to Refugee 1 (2012), who himself fled Imizamo Yethu in 2009:

> the ANC leaders came in on the Sunday and told everyone to calm down. They said we could come back. But it was their friends in SANCO who chased us out the day before. I think they wanted to show the others that were just as much part of the movement as the big townships, but their heart was not really in it.

Importantly, one of the reasons that the xenophobic attacks in Imizamo Yethu of 2008 and 2009 were relatively muted was due to the work of PASSOP (People Against Suffering, Oppression and Poverty), a refugee rights NGO. Through running various programmes and events in Imizamo Yethu, PASSOP had succeeded in generating some recognition of migrants by ANC/SANCO leaders as at least partly legitimate residents, and so dampened the attacks of 2008 and 2009. According to former PASSOP director, Braam Hanekom (2012), 'one of the main goals was to work with everyone in the community to unite all poor people in their common struggle on whatever the issue was. Whenever there was a march, we were there. We talked about rights for all'. PASSOP's presence in Imizamo Yethu was cemented by an advice office where local residents could go for help with various issues, especially accessing state documents, and by the fact that one of its staff, Refugee 1 (2012), lived there and had participated in community structures for 7 years (see Figure 6.5). He says:

> I was well known in Imizamo Yethu. We spoke with the leaders, we spoke with the youth. We spoke with everyone. We would talk about rights, we would explain why people were refugees ... We would help calm issues down when conflict arose ... They knew us, and mostly there were good people.

This approach has much in common with the 'slow activism' employed by the TAC and its allies (Robins 2014). In short, 'slow activism' includes the 'politics of the spectacular' (popular mobilisation and protest) in a suite of tactics alongside the use of courts, representation in parliamentary and other policy spaces, engagement in the media and public debate, and alliance formation with sympathetic organisations nationally and internationally. This

Figure 6.5 PASSOP refugee who left Imizamo Yethu after xenophobic attacks

constellation of tactics is embraced in what is envisaged as a long-term struggle for rights through engaging in both policy work in the public realm and in citizenship building through community mobilisation. It is clear from our own interactions with SANCO and the ANC leadership in Imizamo Yethu that they are very aware of issues of refugee rights and xenophobia, and hold PASSOP leaders in high esteem. It does seem likely that this activism helped reduce xenophobic violence, but it was not enough to prevent it. Nevertheless, this politics represents a rights-based alternative to the avoidance of the state discussed in respect of informal housing (Bayat 2000), and the Chatterjee-like (2004) practice of community representatives acting as gatekeepers for developmental governance noted in the section above.

While life returned to normal fairly quickly after the attacks of 2008, Refugee 1 (2012) reports being disturbed by the events to the point that when a similar event happened in 2009 he decided to leave Imizamo Yethu for good. 'It was sad for me to leave, I had lived there a long time, but the way people turned on me reminded me of bad experiences back home in Zimbabwe, and I felt I should get to a safer place'. The 2009 incident was the destruction and looting of a shack belonging to a Malawian man living very close to Refugee 1. He adds (2012):

> he was at work ... and then this crowd of women gathers and starts accusing him of raping a child. People got very angry and trashed his shack and took his stuff. Later I found out that his girlfriend, who was South African, was among them ... Anyhow, when he came back from work he heard about this and went straight to the police station for protection. No one filed any charges but he never came back. That night they started saying all foreigners must leave IY. I thought how can they do this

again? That man did nothing. And even if he did, why must all foreigners leave? I decided enough was enough.

A critical insight into the 2009 attack is the fact that, perhaps unlike 2008, it clearly had a popular basis to it. Although only a limited proportion of the South Africans living in Imizamo Yethu participated in the attack and openly endorsed anti-foreigner prejudice, it is clear that a significant number do harbour prejudice against foreigners. Indeed, from focus groups we conducted with a variety of foreign residents in 2012, it is clear that belonging is conditional or limited for foreign migrants. Thus respondents described SANCO as 'a South African thing' in which they do not participate (Residents 2012). Part of the reason for this may be choice by foreigners, as one resident explained. However, another part of the reason is that SANCO meetings are conducted in isiXhosa, a fact that has also annoyed some Afrikaans-speaking coloured residents of Imizamo Yethu. When confronted by one SANCO leader with the reply 'but I know you understand Xhosa anyway', the retort was 'that's not the point – you make it hard for us'. According to one young Zimbabwean man (Residents 2012):

> we know we are not fully welcome here ... we are not seen as the same ... so I make sure I am friends with all the gangsters, then nobody messes with me just because I am a foreigner ... you don't want a fight with someone over a girlfriend to become a foreigner thing ... that happens a lot ... foreigners all get blamed when two people have a fight.

From our engagement with SANCO and ANC leaders over many years it is clear they are sensitive about allegations of xenophobia, and are quick to downplay any deliberately xenophobic agendas in Imizamo Yethu. While the relative lack of xenophobic violence in Imizamo Yethu largely bears out these views, in part due to the work of PASSOP, it is also clear that most foreigners we engaged with in Imizamo Yethu feel heightened vulnerability because of xenophobic prejudice. There is thus an important gap between a lack of violence or active discrimination, and equal belonging in a place. Key here is the notion of the party-society, which casts belonging in racial and national terms, effectively excluding foreign African migrants from full membership in Imizamo Yethu. Thus informality has a further aspect beyond the exclusion of the poor, it is also the home of the foreigner. In democratic terms, the informal and foreigners relate to the state as illegal and outsiders respectively, and as such, are excluded from formal democracy.

Conclusion

This chapter has framed an analysis of settlement upgrading in Imizamo Yethu in terms of both the contending logics of developmental governance and informality, and in terms of contests over being and belonging in Hout Bay that take racial, party and national lines. The contests also shed light on the limits of collective control of residents over both formal and informal attempts to upgrade Imizamo Yethu. Thus, the battle over the 'green belt' is a racialised conflict over where and how black migrants should live in Hout Bay. The protest over the Disa School is the manifestation of contest over the representational opportunities and patronage benefits offered by developmental governance, even within a context of party capture. The protest over super-blocking is in significant part informed by a desire to avoid developmental governance, and in particular to protect the informality of Dontse Yakhe from perceived threats of eviction, loss of property, and loss of livelihoods.

Considered together, these two analytical frames of settlement upgrading – governance and belonging – help us understand the weakness of community leadership in Imizamo Yethu. Thus, both developmental governance and informality are disconnected from forms of local democracy, albeit in different ways. While developmental governance in Imizamo Yethu tends towards party capture, and therefore Chatterjee's (2004) 'political-society' with forms of patronage, nepotism and even corruption, informal practices are closer to Bayat's (2000) account of the defence of quiet encroachment. In addition, the racial contest over the vision for the 'green belt' has provided a basis for a factional contest within SANCO; while the partisan and nationalist logic of community representation sits in tension with the multiple nationalities that constitute Imizamo Yethu, further weakening community organisation and leadership legitimacy. This intersection of how governance happens with divergent conceptions of belonging weakens local order, and thus the legitimacy of any one set of leaders.

Notes

1 The 2007 SANCO leader refers to the individual elected as SANCO chairperson in 2007.
2 'Sweat equity' means the contribution to a project in the form of work or labour time rather than money or other material resources.

References

Adriaanse, D. (2017). Imizamo Yethu residents opposed to city's super-blocking. *Independent Online*, 15 August. www.iol.co.za/capetimes/news/imizamo-yethu-residents-opposed-to-citys-super-blocking-10808912. Accessed 5 September 2017.

Ball, S. (2010). An open letter from Su Ball to Cllr Hayward. *Hout & About*, Hout Bay Residents' and Ratepayers' Association, February. www.houtbay.org.za/RAHB_Newsletters.html. Accessed 11 November 2016.

Bayat, A. (2000). From 'Dangerous Classes' to 'Quiet Rebels': Politics of the Urban Subaltern in the Global South. *International Sociology*, *15*(3), 533–557.

CCT (City of Cape Town). (2017a). Consolidated relief efforts underway in Imizamo Yethu. Statement by the City's' Acting Executive Mayor, Alderman Ian Neilson, 13 March. www.capetown.gov.za/Media-and-news/Consolidated%20relief%20efforts%20under%20way%20in%20Imizamo%20Yethu. Accessed 5 September 2017.

CCT. (2017b). Out of the ashes comes a far better Imizamo Yethu, 21 June. www. capetown.gov.za/Media-and-news/Out%20of%20the%20ashes%20comes%20 a%20far%20better%20Imizamo%20Yethu. Accessed 5 September 2017.

Chatterjee, P. (2004). *The Politics of the Governed: Reflections on Popular Politics in Most of the World.* New York: Columbia University Press.

Cronje, J. (2017). One dead, buildings torched in night of Hout Bay protests. *News24*, 22 July. www.news24.com/SouthAfrica/News/one-dead-buildings-torched-in-night-of-hout-bay-protests-20170722. Accessed 5 September 2017.

De Villiers, J. (2017). Misinformation fuelled violent protest in Imizamo Yethu – community worker. *News24*, 21 July 2017. www.news24.com/SouthAfrica/News/ misinformation-fuelled-violent-protest-in-imizamo-yethu-community-worker-20170721. Accessed 5 September 2017.

Harte, W., Hastings, P., & Childs, I. (2006). Community Politics: A Factor Eroding Hazard Resilience in a Disadvantaged Community, Imizamo Yethu, South Africa. Paper presented at the Social Change in the 21st Century Conference, Centre for Social Change Research, Queensland University of Technology. http://eprints.qut. edu.au/6132/. Accessed 7 September 2013.

Hout Bay & Llandudno Environment Conservation Group v Minister of Local Government, Environmental Affairs & Development Planning, Western Cape and Others (23827/2010) [2012] ZAWCHC 22. www.saflii.org/za/cases/ZAWCHC/2012/ 22.pdf. Accessed 10 July 2018.

MacGregor, H., Bucher, N., Durham, C., Falcao, M., Morrissey, J., Silverman, I., Smith, H., & Taylor, A. (2005). Hazard Profile and Vulnerability Assessment for Informal Settlements: An Imizamo Yethu Case Study with special reference to the Experience of Children. Cape Town: DiMP, University of Cape Town, March 2005. www.proventionconsortium.net/themes/default/pdfs/CRA/South_Africa.pdf. Accessed 5 September 2017.

Monaco, S. (2008). Neighbourhood politics in transition. Residents' associations and local government in post-apartheid Cape Town. Doctoral dissertation, Uppsala University. www.diva-portal.org/smash/get/diva2:171377/FULLTEXT01.pdf. Accessed 13 November 2016.

Mukadam, S. (2013). Housing Havoc – Part One. *Finding My Voice: My Journey Through Writing.* https://findinghervoice.wordpress.com/2013/10/02/housing-havoc- part-one/

Piper, L. (2015). From Party-State to Party-Society in South Africa: SANCO and the Informal Politics of Community Representation in Imizamo Yethu, Hout Bay, Cape Town. In Bénit-Gbaffou, C. (Ed.), *Popular Politics in South African Cities: Unpacking Community Participation.* Pretoria: HSRC Press, 21–41.

Piper, L., & Anciano, F. (2015). Party over Outsiders, Centre over Branch: How ANC Dominance Works at the Community Level in South Africa. *Transformation: Critical Perspectives on Southern Africa, 87,* 72–94.

Rangasami, J., & Gird, A. (2007). Rapid Impact Assessment of NMTT's work in Imizamo Yethu, Cape Town, from 2003 to 2005. Report prepared for the Niall Mellon Township Trust. *Impact Consulting: Cape Town.* www.impactconsulting. co.za/downloads/Mellon%20Housing%20Initiative%20Rapid%20Assessment.pdf. Accessed 13 November 2016.

Robins, S. (2014). Slow Activism in Fast Times: Reflections on the Politics of Media Spectacles after Apartheid. *Journal of Southern African Studies, 40*(1), 91–110.

Shaug, E. (2009). When taking the high ground can be a hazard. *Cape Times,* Monday 5 October 2009. www.houtbay.org.za/index.html#IYdev. Accessed 13 November 2016.

Sikota, Z. (2015). No meaningful participation without effective representation: the case of the Niall Mellon Housing Project in Imizamo Yethu, Hout Bay. Master's thesis, University of the Western Cape.

Swimmer, L. (2012a). Len's Lines. *Hout & About*, Hout Bay Residents' and Ratepayers' Association, July. www.houtbay.org.za/RAHB_Newsletters.html. Accessed 13 November 2016.

Swimmer, L. (2012b). Len's Lines. *Hout & About*, Hout Bay Residents' and Ratepayers' Association, September. www.houtbay.org.za/RAHB_Newsletters.html. Accessed 13 November 2016.

Swimmer, L. (2012c). Len's Lines. *Hout & About*, Hout Bay Residents' and Ratepayers' Association, December. www.houtbay.org.za/RAHB_Newsletters.html. Accessed November 2016.

von Lieres, B., & Piper, L. (Eds.). (2014). *Mediated Citizenship: The Informal Politics of Speaking for Citizens in the Global South*. Basingstoke: Palgrave Macmillan.

Yiftachel, O. (2009). Critical Theory and 'Gray Space': Mobilization of the Colonized. *City, 13*(2–3), 246–263.

Zille, H. (2007). 'The true story of Imizamo Yethu', *Cape Times*, 14 February, Edition 1.

Zuern, E. (2011). *The Politics of Necessity: Community Organising and Democracy in South Africa*. Pietermaritzburg: University of KwaZulu-Natal Press.

Interviews

1. ANC leader. (2015). Interviewed by Laurence Piper. 20 May 2015.
2. ANC Youth League leader. (2012). Interviewed by Laurence Piper. 29 May 2012.
3. City official 2. (2015). Interviewed by Fiona Anciano. 30 March 2015.
4. City officials. (2017). Interviewed by Laurence Piper 10 August 2017.
5. Community leader 2. (2012). Interviewed by Laurence Piper. 10 May 2012.
6. Community leader 2. (2017). Interviewed by Laurence Piper. 24 August 2017.
7. Mediator 2. (2017). Interviewed by Laurence Piper. 25 August 2017.
8. Refugee 1. (2012). Interviewed by Laurence Piper. 12 April 2012.
9. Refugee 2. (2012). Interviewed by Laurence Piper. 20 February 2012.
10. Residents. (2012). Workshop on insecurity in Imizamo Yethu with foreign residents. 18 May 2012.

7   Taxis, violence, and leadership in Imizamo Yethu

In April 2015, a protest march took place against the proposed MyCiti Bus Rapid Transport (BRT) station at the northern entrance to Imizamo Yethu (Figure 7.1). Although not confronted by police, tempers rose and the protest culminated in the destruction of eight aluminium-framed temporary houses built close to the informal taxi rank. Given the long-standing demand for housing in the settlement, and the fact that the protest was around transport issues, the destruction of the temporary houses was initially confounding. However, on closer inspection it turns out that these issues were connected. According to the recently elected leader of the ANC-aligned faction of SANCO (hereafter SANCO leader),

> the city could not explain why these few people had got housing ahead of others who have been waiting for years ... also why was MyCiti going where we had agreed there would be housing? This angered the community and we tore the temporary houses down.
>
> (SANCO leader 2015)

In addition to confirming the centrality of the 'battle of the green belt' to politics in Imizamo Yethu, this incident draws attention to issues of transport, violence, and leadership in the settlement. The main argument is that, similar to conflicts over housing outlined in the preceding chapter, conflicts over transport in Imizamo Yethu are rooted in the contradictory logics of contending forms of governance. Unlike in relation to housing, market governance is important to understanding transport in Imizamo Yethu as the minibus taxi industry is a major means of mobility for most residents. However, both informal practices in the form of local sedans, and developmental governance in the form of the Bus Rapid Transport system of the City, are also key to these dynamics.

This chapter demonstrates these tensions between forms of governance in the following ways. We show how the City attempted to license informal taxis in Imizamo Yethu in response to various externalities, in particular the use of violence by taxi owners and drivers to defend their livelihoods. We also show how both informal and formal taxis, driven by market competition, exist in

Figure 7.1 Remains of eight aluminium temporary houses destroyed in anti-BRT protest

tension with public transport. In order to reduce the contradictions between market and developmental goals in the transport sector, the City of Cape Town is attempting to introduce a BRT system that includes formal taxi owners affected by the new bus routes as owners in the new BRT Company. In seeking to provide safer, more affordable, and environmentally friendly transport to all residents of Cape Town, but on a for-profit basis, the BRT is a hybrid of developmental and market governance. As we shall see, however, while this public–private blend reduces tensions with most formal taxi owners, and thus the threat of violence and protest, it does not eliminate them altogether.

Further, conflicts over transport produce legitimacy problems for leaders of Imizamo Yethu. This is because SANCO's primary role is to mediate relations to the state, in particular to mediate forms of developmental governance, but as transport is governed through formal market and informal relations, there is little to no role for SANCO in this process. In short then, as informality and market relations increasingly dominate the supply of key needs such as transport to residents of Imizamo Yethu, SANCO, in any faction, becomes less relevant to the governance of daily life.

Lastly, despite being a settlement of substantial insecurity and violence, civil or political violence has been remarkably rare in Imizamo Yethu – at least until the aftermath of the fire of 2017 described in Chapter 6. This we take as a sign of the enduring legitimacy of the ANC's right to rule in the settlement, which is at the heart of the 'party-society' idea. At the same time however, disconnected from informal and market governance that shape how people get around, riven by factionalism and associated with corruption and nepotism, the community leaders aligned with the ANC are in a weaker position today than ever before. Thus, both increased violence in the taxi industry and decreasing legitimacy for community leaders are products of forms of governance outside of City, and therefore democratic, control.

Taxis and the 'battle of the green belt' in Imizamo Yethu

By 2015, the ANC-aligned faction of SANCO had elected a new leader who was a key mover in the anti-BRT protest described above. We spoke to him a few months after the protest, and toured the site where the temporary houses lay destroyed. When asked why he did not take the bent aluminium frames for recycling, he replied, 'no, no, no, we can't do that, it belongs to the City' (SANCO leader 2015). In rejecting the opportunity to recycle the aluminium frames destroyed in the anti-Bus Rapid Transport protest, the SANCO 2015 leader demonstrated a clear understanding of the difference between an action that is collective and politically motivated and one that is individual and economically self-interested. Thus, it is one thing for a community protest to destroy houses that, ironically, symbolise the long-standing failure of the state to deliver houses to the poor, but another for individuals to make personal profit out of this event.

That the protest against the BRT turnaround station was really an extension of the 'battle of the green belt' is confirmed not just by the SANCO leader (2015), but also by Bristow's (2015) research. At a public meeting on the proposed BRT turnaround in November 2014, attended by the MyCiti implementation department, the Housing Department, SANCO, and various taxi owners, the mood of the gathering was angry (Bristow 2015: 70). From the outset the debate centred on issues of housing, and especially the temporary housing, rather than the BRT. Key issues raised included: that moving eight families from the proposed site gave them unfair preference over others who had been waiting longer; that moving these eight shacks jeopardised formal construction by encouraging others to put up shacks; and that the new aluminium houses were not even being used by the families but were being rented out to foreigners. After 30 minutes, the meeting had descended into chaos and was disbanded. According to one participant:

> Since 1990 this area is called 16 hectare and was earmarked for development for the people. And then there was infighting by some of the people because they wanted to use the land for their own children. Then there was a court interdict and then it was removed. Now they say that they can start building houses there. They start developing for the bus instead

of the houses. So we have questions again, why? Because they are now prioritising the bus that came yesterday and we don't know because we have been here for years and the bus came yesterday they did not cater for us.

(Bristow 2015: 70)

Another added, 'we are looking to toyi-toyi, to burn tyres. But it is a secret. To chase MyCiti because they don't talk to us' (Bristow 2015: 67). A few months after the meeting, the aluminium houses intended for those to be displaced by the new BRT turnaround station were trashed.

There is more to the story than this, as there is more to the SANCO leader than is first apparent. As well as being a community leader, the SANCO leader is also a taxi owner, and one of several in the Hout Bay Cape Town (HBCT) Association that has not been included in the City of Cape Town's BRT buy-out scheme for taxi owners. Consequently, his business is under direct threat from the MyCiti buses scheduled to operate on the Wynberg Route that runs on Hout Bay Road past Imizamo Yethu. This opens up the possibility that, despite its communal camouflage, the anti-BRT protest was really driven by the business interests of an entrepreneur. Ironically, this is precisely the moral distinction between the communal, political, and legitimate, and the personal, economic, and illegitimate identified above by the very same SANCO leader.

However, this conclusion is premature and partial for several reasons. First, there is the long-standing history of struggle over the development vision for Imizamo Yethu, especially over the 16 hectares that lie along Hout Bay Road at the lowest edge of the settlement. Thus, even if the SANCO leader has a personal interest in delaying the BRT on the Wynberg route, it coincides with a long-standing collective demand for housing on the BRT land. In this sense it constitutes a classic instance of von Holdt's 'dual nature of protest' where elite interests and popular insurgency combine in one event (von Holdt & Alexander 2012: 104). Second, the advent of the BRT threatens the livelihoods of at least some taxi operators on the Wynberg route who have a case that their exclusion is unjust – and again framed in terms of racial exclusion. To understand this better it is helpful to know the history of formal taxis in Hout Bay.

Formal taxis, the BRT, and economic exclusion

Hout Bay is connected to the rest of Cape Town by three roads, one of which is the Chapman's Peak toll road that leads south to the Cape peninsula. As discussed in the subsequent chapter, this is an expensive toll road and few people commute south for work, so most traffic flows either along the Atlantic seaboard to the city centre or through the leafy green suburb of Constantia to Wynberg. The rapid growth of the population of Hout Bay in the last 20 years, and especially Imizamo Yethu, has created a significant need for public transport as the vast majority of poor residents do not own cars.

Research by the Sustainable Livelihoods Foundation found that by 2013 this need for transport had yielded four different taxi industries (see Figures 7.2 and 7.3). Two of these are formal, involving over 65 minibus taxis registered with CATA (Cape Amalgamated Taxi Association) travelling on the Hout Bay to Wynberg route, and over 200 CUTA (Cape United Taxi Association) taxis using the Hout Bay to City Centre route on the Atlantic seaboard. In addition, two informal taxi groupings transport residents from Imizamo Yethu around Hout Bay. These are colloquially known as the 'amaphela' (cockroaches), and in 2013 were in the process of being formalised into the Hout Bay to Cape Town (HBCT) Association, as well as the 'amahoender' (chickens), which at that time were not formalised. Where most residents of Imizamo Yethu use formal taxis to travel to work or further afield, the informal taxis are used

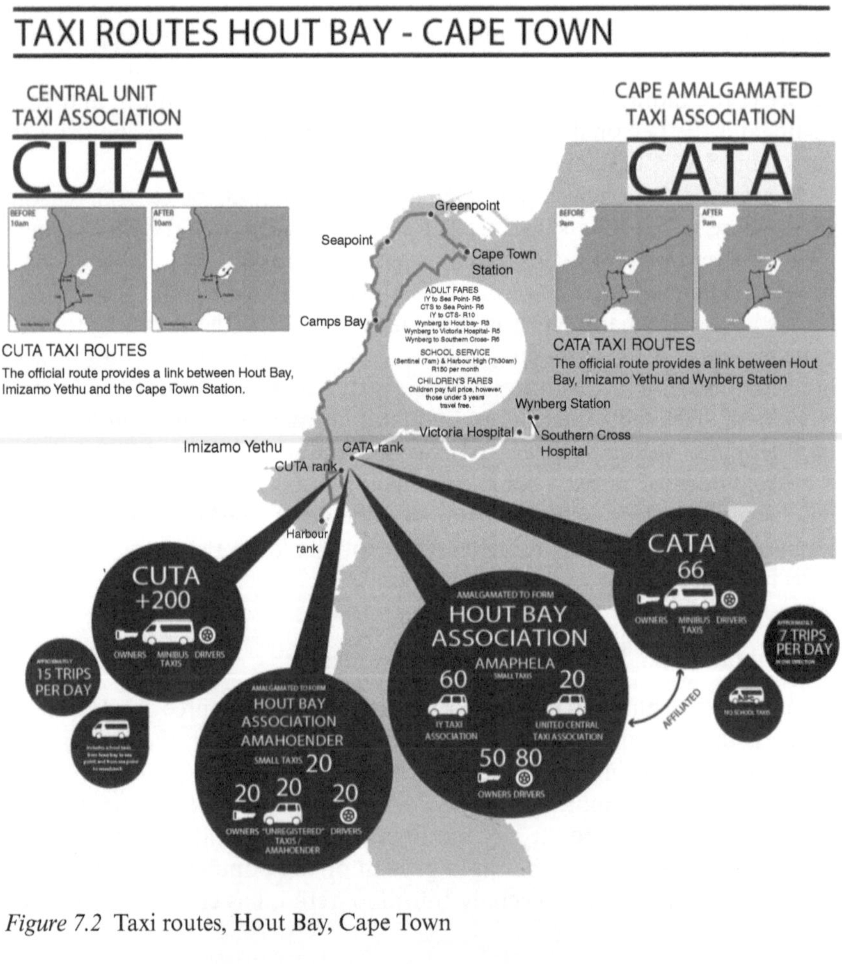

Figure 7.2 Taxi routes, Hout Bay, Cape Town

LOCAL ROUTES

Informal taxis, known as 'amaphela' (meaning 'cockroach') and 'amahoender' (meaning 'chicken') provide short-distance transport services. Informal taxis provide workers with transport from Imizamo Yethu to their work destinations in Hout Bay and Llandudno. They also provide the township residents with access to Hout Bay shopping malls, the clinic, and other facilities. The majority of the informal taxi drivers live in Imizamo Yethu, and the transport service is predominantly owner-operated.

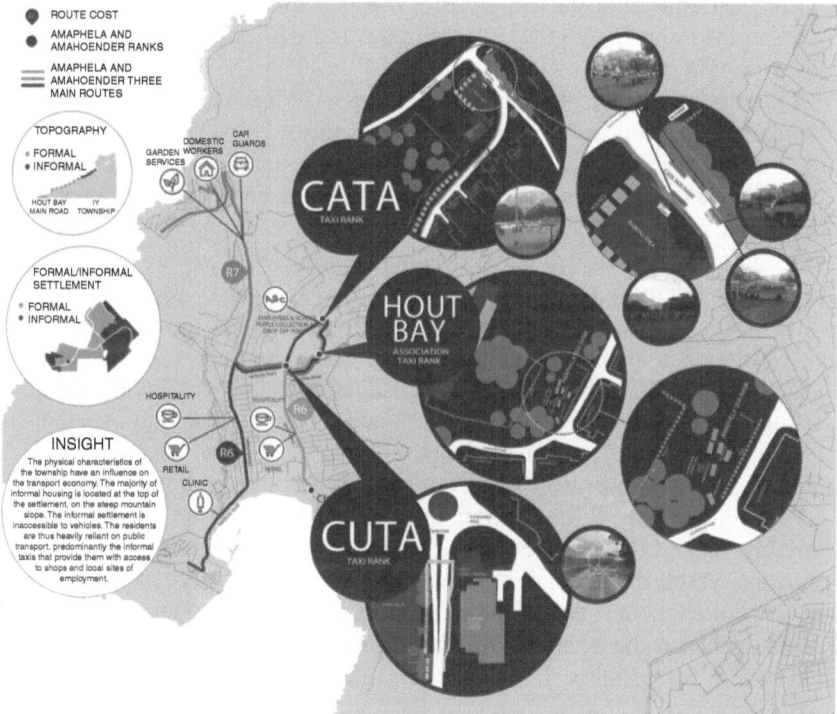

Figure 7.3 Local routes

mostly to go to the local shops, and are especially popular when carrying a load of groceries up the steep slopes of Imizamo Yethu.

As also illustrated by Figure 7.2, the implicit business model of taxis is low profit per fare, but high volume of fares, with the taxi driver first meeting daily targets and then earning their own wage. In addition to incentivising the transport of large volumes of passengers, this model also transfers much of the risk associated with petrol price increases onto the driver first. The taxi industry is thus not just competitive, but this competition most keenly affects drivers whose livelihoods depend on moving large numbers of people quickly (Taxi drivers 2013).

In this regard, it is critical to note that the taxi industry mushroomed in the late 1980s in response to deregulation but routes quickly became over-saturated, and competition for customers degenerated into incidents of conflict (McCaul 1990; Barrett 2003). Largely unregulated, taxi operators resorted to violence to defend and secure their livelihoods, exacerbated by corrupt relations with officials and the rise of mafia-like 'mother bodies' (Dugard 2001). This conflict led to the tightening of regulations on taxis, including the issuing of licences for specific associations on specific routes to manage conflict. Thus, individual taxi owners would apply to join an association that enjoyed exclusive rights to a route. In Hout Bay, CUTA has the rights to the Hout Bay to City Centre route, and CATA the rights to the Hout Bay to Wynberg route.

This model of reducing open competition to reduce violence has informed the taxi industry up to the present. As we shall see below, a similar logic was evident in our experience of informal taxis in Imizamo Yethu, as competition over the right to operate on crowded routes descended in violent attacks. This provoked a similar response from the state to formalise these through licensing to provide a degree of livelihoods security. Thus, as with housing policy that attempts to reduce the health, fire and safety costs of informal settlement, formalisation of taxis attempts to reduce the externalities (violence, safety, bad driving) of informal market relations.

At the time of the BRT protest in 2015, tensions between the formal associations along the two main routes remained. There are two main reasons for this. First, was the allocation of the Hout Bay to City Centre route to CUTA, most of whose owners were not from Hout Bay, and which marginalised the CATA-aligned HBCT Association, most of whom were from Imizamo Yethu. Notably, these CATA owners had to register at the Wynberg taxi rank, as Imizamo Yethu did not have its own taxi association in 2013. Second was the introduction of the BRT system to Hout Bay that does not recognise CATA rights to the Atlantic route, and threatens many in the taxi industry that have existing rights. This is because full compensation for routes lost by taxi operators and shares in the new BRT Company go to those whose entire route is supplanted by the BRT. This does not apply to many taxi owners in Hout Bay, and partially affected owners get only partial compensation and no shares.

Competition over routes

In respect of the first issue, CATA-aligned taxis operators claimed to be the original group who started the Hout Bay to Cape Town route along the Atlantic seaboard; their rights to use the route were, however, not formalised. Indeed, as one respondent observed, 'Imizamo Yethu is the only township without a taxi association' (Taxi owner 2012). In the absence of a formalised presence and rights to the route, competition from CUTA-aligned taxis quickly emerged. According to one respondent (quoted in Bristow 2015: 69):

> It was a fight between us and CUTA. We told them that they could not operate that route and that it was ours. But to be honest we did not have

mini-busses just the sedans ... So we had this fight and the traffic cops came here and operated for a week impounding our vehicles. So then we decided that we must do this right.

While the CATA taxi drivers decided to withdraw from operating, the CUTA owners managed to be the first to secure licences for the route. How exactly they managed to do this is a matter of contention. According to the licensing authority, the Provincial Regulatory Unit (PRE), CUTA registered in 2003 and applied for rights shortly thereafter. Notably, they add that

> the PRE does not necessarily determine which association to be registered when and where but rather we receive applications from associations. These applications will then be referred to the Planning Authority in the area where the association is intended to provide the service.
>
> (Bristow 2015: 70)

In short, although formally a matter for the province to approve, someone in the City of Cape Town affirmed the right of CUTA to the lucrative Atlantic route. Whatever exactly happened in this matter remains mysterious, but the incident is a good example of both the limited power of the City in the context of co-operative governance in South Africa, and the added complexity this lends to bureaucratic governance by the state.

The CATA-aligned taxis, now renamed the Hout Bay Cape Town (HBCT) Association, began contesting CUTA rights to the Atlantic seaboard route in 2004. According to Bristow (2015: 68), one document from the City of Cape Town states that, contrary to the PRE decision, CATA's claim to the Atlantic seaboard route was supported by the City. She notes, 'officials from the South Peninsula Traffic Department had visited Imizamo Yethu and established that there was a need for transportation from the informal settlement to Cape Town via Camps Bay'. However, the Association could not operate until their licence had been approved, and as the Public Protector Report notes (Bristow 2015), while the HBCT Association was waiting for the licensing documentation to be approved, CUTA started to operate on the same route. Following the use of the courts, the HBCT eventually managed to register with the PRE in December 2012, but by this point events had been overtaken by the BRT process.

The rise of the bus

The introduction of the BRT system has had a profound impact on the taxi industry in Hout Bay. Informed by the revolutionary public transport system developed in Bogota, Columbia, the BRT was conceived as a key part of an integrated public transport network to improve upon the existing public transport in the build-up to the 2010 Soccer World Cup. Confronted by the spatial legacy of apartheid where poor, black residents still live a long way from work, the BRT attempts to address this by making transport more affordable, but also safer, more environmentally friendly, and accessible to the disabled and elderly. It thus pursues clear developmental ends. At the same time, the key business feature of the BRT is to buy out taxi owners by incorporating those with licences into vehicle operating companies that would be contracted by the City of Cape Town to run the MyCiti buses for 12 years. The key idea was to give taxi owners a vested interest in the BRT and so promote co-operation rather than conflict. By and large this approach appears to have worked. Indeed, according to a City councillor, the Hout Bay route is 'the top performing route or the second performing route' (Bristow 2015: 70). In general however, the BRT makes a loss. In 2016, the MyCiti buses in Cape Town recovered just 49 per cent of their operating costs from fare revenues (Van Rensburg 2017).

However, in Hout Bay, three issues remain with the introduction of the BRT. The first is that CUTA was bought out along the Atlantic route by the BRT process and not CATA, despite CATA's reasonable claims to have rights to this route. This issue continues to irk the CATA-aligned taxi owners. Their anger is evident in the comments of one taxi driver who stated 'Those busses! One of these days, they force and put people in them, we will burn those busses if we have to and will go to jail if we have to. We will fight the City of Cape Town down to the wire' (Bristow 2015: 67).

The second issue is that not all taxi owners working the Hout Bay to Cape Town route were included in the new deal as they were only partially affected by the BRT. Further, a number of taxi drivers summarily lost their jobs as their owners bought into the BRT without informing them. As a consequence, a rump of CUTA taxi owners continued to operate illegally in Hout Bay, using an informal taxi rank near the police station for their activities (Bristow 2015: 54; Taxi owners 2012). As one driver put it:

> We are on our own fight to get permits. We came here at a very young age; we were all gaartjies (conductors) before we became drivers. Now all of a sudden the owners are turning their permits in. Without even letting us know. We all had dreams of owning our own taxis one day. They are turning their permits in and they are closing our route. So it is bad for us.

Lastly, although the MyCiti buses were not yet running on the Hout Bay to Wynberg route in 2012–2013, several taxi owners who work that route were deeply concerned about the threat to their business posed by the new buses (Taxi owners 2012). Excluded from the Atlantic route deal, and witnessing the marginalisation of partially affected CUTA operators, CATA taxi owners are loath to trust the city. This distrust has deep roots linked to contests over local leadership, party loyalty, and accessing the City.

Informal taxis and 'party-society' in Imizamo Yethu

Clearly, the taxi industry in Hout Bay is very competitive, but this competition is manifested most sharply for drivers whose livelihoods depend on moving large numbers of people quickly. The intensity of this competition, including among the informal 'amaphela' and 'amahoender', was evident on our first day of fieldwork in Imizamo Yethu in 2011. On entering the settlement we found the main road blocked near the informal taxi rank, and saw at least five vehicles damaged from rocks thrown at them. At the Hout Bay police station we met an 'amahoender' taxi driver reporting this attack who complained that rival 'amaphela' taxi drivers damaged his car. These incidents were the culmination of several days of conflict that reportedly saw one person wounded in a shooting incident (Isaacs 2011).

From engaging with taxi drivers on both sides of the conflict and in both factions of SANCO, we were able to establish that the conflict centred on the attempt to formalise the local taxi industry into what later would become the Hout Bay Cape Town (HBCT) Association affiliated to CATA. The issue was the limited number of licences promised by the City that made the 'amaphela' reluctant to admit new members. Driven by the necessity of earning a livelihood, many (about 20) were not accommodated in the 'amaphela' (about 80), but decided to drive anyway. These renegades were called the 'amahoender' (SANCO 2007, 2012; Taxi owners 2012).

This issue continued to simmer over the next couple of years until the HBCT Association agreed to include most of the 'amahoender' in their ranks, thereby defusing the conflict. Notably, in the early days local leaders articulated a variety of causes of the conflict. Among these were the views of some local leaders that the 2007 faction of SANCO was using this to destabilise Imizamo Yethu and therefore delegitimise the 'new' ANC-aligned faction. Another was that the DA government was using the violence to discredit taxis, so as to bring in the BRT system (Taxi owners 2012). While in retrospect these theories seem somewhat unlikely, they do confirm the centrality of race, party, and the relation to the local state as central to conceptions of politics in Imizamo Yethu.

Hopefully, it is now clear how issues related to taxi violence (the BRT protest and amahoender versus amaphela conflict), reflect the long history of partisan conflict over the development of Imizamo Yethu. Key to this are not just divergent visions for Imizamo Yethu, but also partisan visions of how state and society should relate in this process. Thus, the ANC and its allies at local level advance a conception of state–society relations that we term 'party-society' politics (Piper 2015; Piper & Anciano 2015). Central here is the idea of ANC dominance of the state, at least until recently, giving rise to the idea of a 'party-state' akin practically to a one-party or dominant party system (Giliomee & Simkins 1999; Brooks 2004; Butler 2009). We suggest that a similar relationship between the ANC and civil society exists in the mind of most party activists.

The attempt to constitute 'party-society' applies at both national and local level. Thus the ANC is, or at least was, able to extend its authority over civil society by forming the tripartite alliance with the Congress of South African Trade Unions (COSATU) and the South African Community Party (SACP). This occurs informally at local or community level through a combination of ideology and patronage politics (Piper 2015; Piper & Anciano 2015). On the one hand the ANC's liberation nationalism includes a 'bounded pluralism' that practically affirms the black oppressed as at the heart of the nation, and a conception of leadership that exclusively constructs the ANC as the only legitimate representative of this implicitly racialised nation. On the other hand, access to the ANC-run local state is best secured through informal networks in the party and its allies, such as SANCO, rather than through formal channels or invited spaces of participatory governance (Bénit-Gbaffou 2011).

In Imizamo Yethu the power of the idea of 'party-society' as the champions of the settlement is evident in the discourse of local leaders. Interviewed in 2015, a SANCO leader stated, 'We as ... the ANC, the people ... we fight to get this piece of land. We were holding the ANC flag from these five informal settlements ...' (Sikota 2015: 65). Also important is the way that SANCO explicitly identifies with the ANC and campaigns for it at election time. Critically, the justification for the ANC is always linked to notions of race and the exclusion of residents of Imizamo Yethu from Hout Bay. A good example is from a focus group with ANC members (ANC voters 2015):

> I used to hear about Hout Bay while I was still in the Eastern Cape and I came here finally and worked as a domestic worker. We used to get arrested here for just being black. We couldn't even build our shacks. What you see now here you would never find during the times of apartheid. We used to stay in the bushes covering ourselves with shelters made of plastic during the night and hiding them away during the day when we went to work so that the police could not see that there were people living there.
>
> The first time I got arrested I was pregnant with my first child. That's the kind of life we lived and by that time Tata Mandela was behind bars, busy fighting for us and he never gave in to the demands and bribes of white people. During this time people were dying and some of us got badly injured. But finally he got out and we won.
>
> Coming to your question, and looking at how much we suffered to get here, I was not expecting to see anyone especially here in Hout Bay where we fought so hard. I was not expecting to hear people saying 'I belong to DA' or 'I belong to Agang' or EFF for that matter. What I was expecting is that we would unite and speak with one voice because we know where we come from. Some comrades died fighting for us. So in a place as small as Imizamo Yethu I was not expecting anyone to be belonging to any other party, it is exactly like selling each other out!

In addition to the influence of racial ideas on ANC legitimacy, in Imizamo Yethu the logic of the 'party-society' is reproduced instrumentally by developmental governance that requires representatives from communities 'below' or smaller than the ward for various development projects such as housing, schools, and clinics. Thus, while the formal political system does not recognise representation below the ward level, developmental governance often requires it at the level of place, offering a space for the 'party-society' to reproduce itself. In the previous chapter, we demonstrated how SANCO was able to do this at the expense of Sinethemba through the governance opportunities provided by various housing projects. This point was made many times by an important SANCO leader who said, 'all I want is development for Imizamo Yethu' (Community leader 2 2015). It is most evident in the fact that the conflict between the two factions of SANCO plays out in terms of who speaks for Imizamo Yethu in development projects from Niall Mellon housing, to the Disa School, to the new clinic and the new security patrols.

While developmental governance is the focus of the representational ambitions of the SANCO factions, these groupings also play an intermediary role between local residents and important aspects of bureaucratic governance too. Thus, while SANCO is not necessarily the first port of call for residents when dealing with issues managed by the line departments of the city such as roads, sanitation, electricity, and the like – residents can access the local state directly on these issues – local leaders are mediators for access to bureaucracy in at least three other ways.

First, a key leader in this network is one of two Community Development Workers (CDW) for Hout Bay whose job it is to help citizens access state services. Very important here is the role of the CDW in facilitating access to short-term state jobs through the Extended Public Works Programme (EPWP) and similar programmes. Second, the ANC is key in respect of authenticating someone as a resident of Imizamo Yethu, a requirement for bank accounts, cell-phone contracts and so on. This residential requirement poses a challenge for residents who do not receive accounts via the mail or, as in the case of informal settlements, have no formal address. As was confirmed by multiple observations, most institutions in Hout Bay, including for example the local Standard Bank, accept proof of residence in Imizamo Yethu on a state form with an ANC stamp on it.

Third, before it was burnt down, all of these functions were administered out of the ANC offices in Imizamo Yethu that also contained the only photocopier available to the public in the settlement. Symbolically and practically, then, the ANC-aligned leadership network in Imizamo Yethu has succeeded in making itself a critical intermediary for residents, especially poorer residents, looking to access both developmental and bureaucratic governance in Hout Bay. Indeed, an ANC leader boasted of how a former party youth leader, who tried to leave the ANC and run a non-partisan development effort, soon returned to the party-fold once he was 'cut off from things' (Community leader 2 2015).

Lastly though, as noted in the previous chapter, there is a severe limitation to the practice of 'party-society' politics in Imizamo Yethu, namely the fact that, since 2006, the City of Cape Town has been governed by the ANC's largest opponent, the Democratic Alliance (DA). Further, the City (and the

Province since 2009), embraces a vision for state–society relations that differs radically from the ANC and SANCO. In contrast to the ANC's racialised nationalism, which legitimates its exclusive right to rule, the DA affirms plural conceptions of community and a liberal contest for office that nevertheless allows it to claim state development projects as party achievements (Piper 2015; Anciano 2016). Thus, where SANCO sees DA rule as inherently illegitimate, the DA sees SANCO's representation as inherently suspect. These differences have created a dilemma for SANCO leaders in Imizamo Yethu between loyalty to the ANC and its discourse of racialised liberation nationalism versus embracing the non-partisan, all-inclusive model of community representation preferred by the DA-run City (Piper & Bénit-Gbaffou 2014).

In this context, a set of SANCO leaders emerged in 2007 who chose to participate heavily in the forums of local governance in Hout Bay, embracing the 'non-partisan' model of community representation advocated by the DA controlled city. In the words of their leader 'we are committed to development … we are not like the other SANCO who only want money from houses and do nothing … they call us headless chickens because we don't chase money (SANCO 2007, 2012). By 2011, ANC leadership, concerned by the perceived co-option of the 2007 SANCO by the state represented in choosing the school over housing as outlined in the preceding chapter, endorsed a rival group. The leadership rivalry between two factions of SANCO in Imizamo Yethu is widely observed in Hout Bay, and one reason why SANCO's standing has declined over the last 10 years. Another, noted in the previous chapter, is the widespread perception of corruption and/or nepotism in the access to housing lists alleged against both sets of leaders. A third reason, identified in this chapter is the increasing irrelevance of a leadership defined by its gatekeeping role in developmental and bureaucratic governance in a community where a huge proportion of residents live by market and informal means.

Informal taxis, violence, and 'slow activism'

The conflict between the 'amahoender' and 'amaphela' in 2011 draws our focus not just to informality and leadership in Imizamo Yethu, but to violence too. It is clear that Imizamo Yethu is a violent place and most people live with high levels of insecurity. Police statistics for Hout Bay indicate an average of 13 deaths per year for the last 5 years (SAPS 2015). This amounts to a figure of 26 per 100,000, which, while lower than the national average of 33 per 100,000, is still high. Further research also suggests that many crimes, particularly those related to sexual violence, mugging, and robbery, are significantly under-reported (Piper & Wheeler 2016). Perhaps more striking has been the rise in drug-related crime, which has seen a notable spike in the last few years.

In participatory workshops run in Imizamo Yethu in 2011, all discussions highlighted the question of insecurity. As Piper and Wheeler (2016: 35) note, almost all groups of residents fear crime, 'especially at night, and in all parts of the settlement other than in the section where they lived'. While respondents felt that 'the police, SANCO/ANC and the community' should be the leading actors in reducing crime, in that order, they reported that those who made the community safe were 'cats and dogs, neighbours and family', in that order. Cats, because they 'kill rats and mice that eat food', and dogs because 'they bark at tsotsis' (Piper & Wheeler 2016: 36).

By contrast, until the vigilante killing of gang leaders in 2015, we had not heard of any civic violence-related deaths in the previous 10 years. Of course, there has been violence directed against property, for example, the aluminium-framed shacks destroyed in the BRT protest or the damage to the cars of the 'amahoender' taxi drivers. In the previous chapter we also noted the two moments of xenophobic mobilisation in 2008 and 2009, and the burning of the ANC office and ANC-aligned leaders' houses in 2017. All things considered, however, the extent and nature of civic violence in Imizamo Yethu has been low compared to most townships in an era of unprecedented popular protest that ranks among the highest in the world (Alexander 2010).

This comparative lack of civic violence is important, first, because as Piper and Wheeler (2016) point out, it runs against common conceptions that political competition for office is the primary source of violence, and suggests that violence is not the basis of local authority in Imizamo Yethu. Hence, the deep, enduring, and public differences between the two factions of SANCO have very infrequently manifested in violent conflict, until recently, that is (Community leader 2 2015; SANCO 2007 2012). As noted in the preceding chapter, following the fire of March 2017, new levels of civic violence became evident in Imizamo Yethu with the burning down of the ANC office and the houses of the Community Development Worker and two ANC leaders.

Propagated by residents of the informal settlement in Dontske Yakhe, this violence mirrors the violence of the informal taxi conflict of 2011, and alerts us to the disconnect between SANCO leaders, consciously positioned as gatekeepers of developmental governance, and the growing numbers of informal settlers and informal livelihoods in Imizamo Yethu that look to avoid the gaze of the state. In both these cases violence erupted when the logics of formality and informality collided, rather than in the competition over access to developmental governance. What this new level of violence also reflects is the declining legitimacy of either faction of SANCO, both unable or unwilling to access developmental governance, to represent significant portions of Imizamo Yethu. The logic of the 'party-society' only applies to those who can, or want, to access the state.

In sum then, factionalism, allegations of corruption, and the growth of informality have all weakened SANCO in Imizamo Yethu. Today both factions of SANCO struggle to mobilise more than a few hundred people in a settlement of well over 20,000. Despite this, it is still strong enough to remain the most legitimate civil society formation in Imizamo Yethu through its association with the ANC and its mediatory role with the state and donors. Importantly, the increasing weakness of SANCO does not necessarily imply a decline in popularity for the ANC, as the standing of the former is linked to the importance of developmental governance for residents of Imizamo Yethu, and this is on the decline with the rise of both the market and informality. Conversely, the ANC is an organisation associated with overt politics such as voting and public debate, activities that are episodic rather than everyday for most residents.

All of this implies that, contrary to the decline of SANCO over the last decade, the ANC could remain a powerful contender with the DA in electoral terms. Key here is the promise of the ANC to undo the racial segregation, economic exclusion, and political marginality experienced by residents of Imizamo Yethu on a daily basis, and produced by the contending logics of market, developmental, and informal governance in Hout Bay. Given the enduring and new segregation that has transpired in Hout Bay since 1994, and its overlay with socio-economic and racial differences, it is not hard to see how the link of race to party works for the ANC. Indeed, the same is true for the DA at the local level of Hout Bay. In effect then, electoral competition in a context of neo-apartheid segregation may well reinforce rather than challenge social divides.

Conclusion

In addition to affirming the racialised 'battle of the green belt' as central to the politics of Imizamo Yethu, the anti-BRT protest draws attention to the conflicts over transport, and in particular the formalisation of the taxi industry and its relationship with the BRT system. Key to these conflicts are issues of livelihoods, and how the struggle to survive informal market competition around transport leads to violence, prompting the state to formalise the industry to save lives, reduce conflict, and improve safety.

Furthermore, in an attempt to achieve developmental goals such as safe, affordable, and environmentally sustainable transport, and to avoid further protest and violence, the state has pursued the BRT idea by including taxi owners in the new dispensation through a co-ownership model. In this way market and developmental governance are blended into a public–private partnership to achieve developmental ends, rather than simply introducing a publicly run bus transport system that would generate conflict with private taxis. In effect, then, informality and market governance have shrunk the space for the developmental governance of transport in Imizamo Yethu through the threat of violence.

This shrunken space for the developmental governance of transport, and its relationship to violence, offers insight into the declining role of SANCO leadership in Imizamo Yethu. In the spirit of party-society politics that ideologically melds state, ANC and blackness, and practically monopolises the gateway to the state, SANCO has hitched its legitimacy to developmental governance. However, as more and more residents rely either on informal or market governance to meet their daily needs, SANCO's role becomes less and less relevant, hence SANCO's inability to speak for the residents of Dontse

Yakhe after the fire and its marginality to struggles over taxis and buses in Hout Bay. Indeed, the decline of SANCO is manifested in the lack of violence in the contest for community leadership of Imizamo Yethu. While factionalism and a perception of corruption also taint community leadership, its declining ability to influence key forms of governance helps to explain its waning status.

References

Alexander, P. (2010). Rebellion of the Poor: South Africa's Service Delivery Protests – a Preliminary Analysis. *Review of African Political Economy, 37*(123), 25–40.

Anciano, F. (2016). A Dying Ideal: Non-racialism and Political Parties in Post-Apartheid South Africa. *Journal of Southern African Studies, 42*(2), 195–214.

Barrett, J. (2003). *Organizing in the informal economy: A case study of the minibus taxi industry in South Africa* (No. 993581583402676). International Labour Organization.

Bénit-Gbaffou, C. (2011). 'Up, Close and Personal': How Does Local Democracy Help the Poor Access the State? Stories of Accountability and Clientelism in Johannesburg. *Journal of Asian and African Studies, 46*(5), 453–464.

Bristow, R. (2015). My city or their city? A case study of the Imizamo Yethu Taxi Industry and the MyCiti Bus Services in Hout Bay. Master's thesis, University of the Western Cape.

Brooks, H. (2004). The Dominant Party System: Challenges for South Africa's Second Decade of Democracy. *Journal of African Elections, 3*(2), 121–153.

Butler, A. (2009). Considerations on the Erosion of Party Dominance. *Representation, 45*(2), 159–172.

Dugard, J. (2001). From Low Intensity War to Mafia War: Taxi violence in South Africa (1987–2000). *Violence and Transition Series*, 4 May. www.researchgate. net/profile/Jackie_Dugard/publication/238085248_From_Low_Intensity_ War_to_Mafia_War_Taxi_violence_in_South_Africa_1987_-_2000/links/ 56ab394908acadd1bdccb421.pdf. Accessed October 2017.

Giliomee, H., & Simkins, C. (eds.) (1999). *The Awkward Embrace: One-Party Dominance and Democracy*. Cape Town: Tafelberg.

Isaacs, L. (2011). More taxi violence flares up in Hout Bay *IOL Online*, 13 October. www.iol.co.za/capetimes/news/more-taxi-violence-flares-up-in-hout-bay-1156563. Accessed 8 May 2018.

McCaul, C. (1990). *No Easy Ride: The Rise and Future of the Black Taxi Industry*. Johannesburg: South African Institute of Race Relations.

Piper, L. (2015). From Party-State to Party-Society in South Africa: SANCO and the Informal Politics of Community Representation in Imizamo Yethu, Hout Bay, Cape Town. In Bénit-Gbaffou, C. (Ed.), *Popular Politics in South African Cities: Unpacking Community Participation*. Pretoria: HSRC Press, 21–41.

Piper, L., & Anciano, F. (2015). Party over Outsiders, Centre over Branch: How ANC Dominance Works at the Community Level in South Africa. *Transformation: Critical Perspectives on Southern Africa, 87*, 72–94.

Piper, L., & Bénit-Gbaffou, C. (2014). Mediation and the Contradictions of Representing the Urban Poor in South Africa: The Case of SANCO leaders in Imizamo Yethu in Cape Town, South Africa. In von Lieres, B., & Piper, L. (Eds.),

Mediated Citizenship: The Informal Politics of Speaking for Citizens in the Global South. Basingstoke: Palgrave Macmillan, 25–42.

Piper, L., & Wheeler, J. (2016). Pervasive, but not Politicised: Everyday Violence, Local Rule and Party Popularity in a Cape Town Township. *SA Crime Quarterly*, *55*, 31–40.

SAPS (South African Police Service). (2015). Crime statistics: April 2014 – March 2015. www.saps.gov.za/resource_centre/publications/statistics/crimestats/2015/crime_stats.php. Accessed November 2016.

Sikota, Z. (2015). No meaningful participation without effective representation: the case of the Niall Mellon Housing Project in Imizamo Yethu, Hout Bay. Master's thesis, University of the Western Cape.

Van Rensburg, D. (2017). Bus Rapid Transit bleeding cash. *Fin24*, 26 February. www.fin24.com/Economy/bus-rapid-transit-bleeding-cash-20170226-2. Accessed October 2017.

Von Holdt, K., & Alexander, P. (2012). Collective Violence, Community Protest and Xenophobia. *South African Review of Sociology*, *43*(2), 104–111.

Interviews

1. ANC voters. (2015). Focus group interviewed by MA Candidate, Political Studies, UWC. 27 June 2015.
2. Community leader 2. (2015). Interviewed by Laurence Piper. 20 May 2015.
3. SANCO 2007. (2012). Focus group interviewed by Laurence Piper. 10 May 2012.
4. SANCO leader. (2015). Interviewed by Laurence Piper. 15 May 2015.
5. Taxi drivers. (2013). Focus group by Shannon Royden-Turner. 3 May 2013.
6. Taxi owners. (2012). Focus group by Laurence Piper. 12 September 2012.

8   Protesting Chapman's Peak toll road

Market governance versus environmental politics

On Sunday 5 February 2012, environmental activist Bronwen Lankers-Byrne began a hunger strike against the building of a Toll Plaza on Chapman's Peak Drive. Armed with her protest signs, and consuming nothing but water, she sat at the building site on the mountainside from 7am in the morning until 7pm at night (see Figure 8.1). Lankers-Byrne insisted she would only stop if construction halted and transparent talks were held between civil society groups, the provincial government, and Entilini, the company building the Toll Plaza. Media attention was fierce. On the third or fourth day, Lewis Pugh, the famous endurance swimmer and ocean advocate, joined Lankers-Byrne as a mentor. Before long, however, he started to get involved, even tweeting Western Cape Premier Helen Zille about the Toll Plaza infringing on a World Heritage Site.

About 10 days into the strike, Pugh counselled Lankers-Byrne that government would not concede as they felt it would make them vulnerable to demands for housing higher up the mountains of both Hangberg and Imizamo Yethu. A solution offered itself, however. On Wednesday, 15 February 2012, day 11 of the hunger strike, two other Hout Bay residents, Charlie Gorton and Fiona Hinds, handcuffed themselves to the scaffolding at the building site, and there was a whole day stoppage as a result. Inspired by their example, Lankers-Byrne handcuffed herself to foundations on Monday 20 February, the day that concrete was to be poured. Work stopped immediately. The police asked her three times during the day to leave, but she refused. That evening at 9pm police arrived with a High Court interdict removing her on the grounds of trespassing. Bronwen Lankers-Byrne conceded, and the protest was over.

These dramatic and somewhat surreal events were the climax of protest against the tolling of Chapman's Peak Road in Hout Bay between 2001 and 2012. Despite some fierce resistance, the residents lost the battle, and today every commuter must pay to use one of the three roads in and out of Hout Bay. The protesters did make marginal gains including securing one toll booth rather than two; modifying the design of the toll booth so that it was more visually discreet and compact; and most importantly, providing impetus for the provincial government to renegotiate the contract with Entilini in 2011 in ways that reduced public financial risk. At the end of the day, however,

Figure 8.1 Bronwen Lankers-Byrne on hunger strike at the Chapman's Peak Toll Plaza

provincial government and their business partners overcame local resistance through a combination of tactics including the expensive and complex shield of the law, the finesse of public participation processes, and the co-option of key local leaders.

This case offers important insights into urban governance, including dynamics of democracy and popular politics in Hout Bay and beyond. In terms of governance, the tolling of Chapman's Peak illustrates well how actors external to local government, in this case provincial government and national business, shape local transport. In privatising a public road, the case exemplifies market governance, and key to market governance is the legal contract that simultaneously empowers private actors with the coercive backing of the state, but removes from them the responsibility of accounting to the public. Hence, even the Premier of the Western Cape was unable to undo a contract made by her predecessor against the will of the people it most affects, at least not without paying substantial costs from the public purse.

In terms of democracy, the case illustrates the profound limitations of community power over the places where people live, even under a democratically elected government with requirements for public participation on development projects. In significant part, this is because the institutions of formal democracy have little or no purchase over the decision-makers external to local government, in this case the provincial government and private companies. In addition, this is because the processes of public participation on development projects are consultative rather than binding, and thus objections can be ignored – at least until the point of popular mobilisation that disrupts the project.

Notably, in this case the extent of popular mobilisation was reduced by the innovative use of representative institutions by the Toll Road project management. Key here was the co-option of community leaders from Imizamo Yethu and Hangberg onto the Chapman's Peak Development Trust, allegedly reducing support for the protest. This tactic of inclusion of key leaders in project structures is reflective of a broader trend in developmental politics to interpret 'community participation' as representation. Conversely, for civil society 'community participation' often takes the form of popular mobilisation or protest, as disruption is the greatest weapon possessed by opponents of a development project. Hence, as demonstrated in this case, but also in the dynamics around super-blocking described in Chapter 6, and the politics around the Peace and Mediation Forum outlined in Chapter 9, where democratic relations in developmental governance are weak, participation tends towards either co-option of leaders through representation, or disruption by residents through popular mobilisation.

Lastly, the chapter offers further sets of insights into popular politics in Hout Bay. First, it affirms the importance of environmental issues to the vision for Hout Bay held by many residents of the Valley. Notably, the blend of ecological thought and disruptive politics, which were manifest in the protest, emerged as alternatives to the environmentalist/conservation discourse and legalistic politics of the Hout Bay Ratepayers and Residents Association (HBRRA). Thus, as argued in Chapter 2, while there is clearly green in the politics of the Valley, it comes in different shades. In the case of the Toll Road, there is also a paradoxical relationship between this green politics and the relationship of middle-class people to cars as their main means of transport.

Second, consistent with the argument in Chapter 6, the struggle over the Toll Road foregrounded divergent and racialised conceptions of belonging in Hout Bay. Thus, while green issues resonate with many in the Valley, it is probably fair to say that these are not so important to most residents of Hangberg and Imizamo Yethu. But the larger issue here is the way that some in the Valley invoke conserving the environment to justify opposing both developments like the Toll Road and the in-migration of poor, black residents. In opposing the belonging of black migrants into Hout Bay, this local strain of environmentalism becomes racialised. In contrast, while the ecological view of being in Hout Bay affirms inclusive conceptions of social justice, it is very much a minority view.

Chapman's Peak Drive: privatising risk

Built by press-ganged convicts from 1915 to 1922, long before the era of environmental impact assessments, Chapman's Peak Drive is famous for its beauty and infamous for rockslides and wild fires, many of which have resulted in the loss of human life (see Figure 8.2). Matters came to a head in June 1994, when several cars were trapped on the mountainside by mud- and rockslides, seriously injuring Noel Graham, who was rendered a paraplegic

Figure 8.2 Chapman's Peak drive

by the accident. Dragged unconscious from his car by other motorists, Graham was carried back to Noordhoek amidst falling rocks (Eliot 2013). When he recovered, Noel Graham successfully sued the Cape Metropolitan Council for negligence in failing to close the road prior to the accident. He won the case and was awarded compensation and legal costs, albeit only after the City appealed the decision to the Supreme Court (Cape Metropolitan Council v Graham, 2000). Eventually, Graham won a R4 million judgement (Gosling 2012b).

Although a new management system was introduced after the Graham incident, in 1999 another driver, Ms Lara Callige, was killed by a rock fall in good weather, and the pass was closed by the provincial government, then run by the New National Party (NNP) in coalition with the Democratic Party (DP) (Chapman's Peak website 2016). While the city was deliberating what to do, a huge fire ravaged the peninsula, including the area above Chapman's Peak Drive, causing further rock falls. As noted by the head of the Western Cape Government's Department of Transport and Public Works, by 2000 Chapman's Peak Drive had claimed 13 lives in the preceding 12 years, and many serious injuries (Eliot 2013). The events of early 2000 thus reinforced the need to make the road safer, and to do so urgently.

Based on recommendations from experts, the Western Cape's Transport Branch decided to pursue 'rock barring' – removing loose or dangerous rocks – and contracts were awarded in March 2000 (Chapman's Peak website 2016). However, by May 2000, rock barring was stopped as it became clear this was insufficient to solve the problem. Within four months, by September 2000, the coalition provincial government decided to embark on a public-private partnership (PPP) and proclaim the route a toll road under the Western Cape Provincial Toll Road Act of 1999. Notably, and not coincidentally, Concor put in an unsolicited bid to fix up the road as a toll road in early 2000 (Gosling 2012b), a suggestion that was taken up in what amounted, in government time, to the blink of an eye. A harbinger of what was to follow with the introduction of the e-Toll system in Gauteng by the South African National Roads Agency (SANRAL) 10 years later, the decision to toll Chapman's Peak prompted significant public opposition and protest that dogged the completion of the road for over a decade.

On 21 May 2002, the NNP/DP coalition government concluded a R350 million (or R160 million[1]) agreement with Entilini, whose major shareholder was Concor, for the financing, planning, designing, and rehabilitation of the road, as well as the operation, management, and control of the toll road for the next 30 years (Dreyer, Breytenbach, Watters, Van Oudenhove, & Parring 2005; Gosling 2012a, 2012b; Residents Association of Hout Bay and Others v Entilini and Others, 2012). As Dreyer et al. (2005: 1) note, this was the first toll road concluded directly with a province, the first concluded under the Public Finance Management Act, and the first subsidised toll road. Entilini Concession was a special-purpose company established by the consortium of Concor Holdings, Haw & Inglis, and Marib Holdings. More on Marib below.

Chapman's Peak Drive was formally declared a toll road on 30 September 2002. The main reason given for this was the cost of fixing the road. As the Western Cape Government (2012) stated: 'Without tolling, Chapman's Peak Drive will be permanently closed'. The idea was that a significant amount of the capital to upgrade the road would come from the private sector, and that users would pay for the maintenance of the new system. This reason for tolling was confirmed by the current Premier of the Western Cape, Helen Zille (2017), who stated that, 'maintaining Chapman Peak Drive alone was costing us ten percent of our road and maintenance budget in the province at the time'.

Certainly, the engineering on Chapman's Peak Drive was sophisticated. According to Entilini this was the first road in South Africa to use 'Swiss engineered catch fences', and had 'innovative engineering' in the form of 'concrete cantilever systems and a half tunnel preventing rocks falling into the road' (Dreyer et al. 2005: 2). The primary protection mechanisms introduced were nearly 1600m of 'flexible, high-energy absorbing catch fences' (Dreyer et al. 2005: 4). Two concrete canopies were also built, one with cantilevers and the other on round poles. In addition, the road was widened, and many of the walls were reinforced.

However, opponents of tolling suggest reasons other than safety and the associated engineering costs were behind the decision to toll Chapman's Peak. Most often cited were the profits to be made by the private company and its cronies in government, who imagined large numbers of wealthy foreign tourists flocking to Hout Bay. Hence, in the wrangling over costs that followed, opponents of the road suggested that private money was not forthcoming to the extent suggested, in addition to the claim that public money was being wasted (Environmental activist 1 2017; Environmental activist 2 2014, 2015; Swimmer 2012). Before exploring this, however, it is critical to note how provincial government and Concor were able to ignore the opposition to the Toll Road in Hout Bay through the invited spaces of public participation.

The limits of public participation

Following the announcement of the Toll Road in 2002, many local residents of Hout Bay voiced opposition to the tolling in the media and in the public participation session that followed. Three main reasons were advanced for opposing the Toll Road. First was the cost of the toll, which was prohibitive for many poorer commuters (in 2015 it was R80 for a once-off return transit, and at best R15 for a return for a daily frequent user). Second was the fact, articulated by environmentalists and conservationists, that the upgrades were infringing on a World Heritage Site. Third was the alleged waste of taxpayers' money due to the excessive costs of the upgrades (Environmental activist 2 2014, 2015; Lawyer 2017; HBRRA leader 1 2017).

Legal opposition to the Toll Road was led mostly by the Hout Bay Residents and Ratepayers Association (HBRRA), which challenged the Toll Road in the courts of public opinion and law, and later by Bronwen Lankers-Byrne, an environmentalist, who played a leading role in the protests and direct action against the Toll Road. Nevertheless, this early opposition was overcome relatively easily through three main tactics: co-option of key community leaders on to project structures; ignoring the substantive views expressed in the public participation process; and the costly and complex process of legal challenge.

Co-opting representatives from Imizamo Yethu and Hangberg

While a significant number of Hout Bay residents did oppose the Chapman's Peak Toll Road, with 10,000 signing the petition in 2012 and around 2,500 marching in protest, most were white residents from the Valley. Only small numbers of black and coloured residents from Imizamo Yethu and Hangberg participated in these events. One reason for this is the limited resonance of green issues in these communities. This is especially the case when the environment is framed as in opposition to development. Further, in the case of Hangberg, the current leader of the PMF claimed they were not engaged to join the protest (PMF leader 2017); in the case of Imizamo Yethu the organisers of the protest approached the 2007 SANCO leadership who mobilised just a handful of people. They did not engage the ANC-aligned faction of SANCO (Community leader 2 2015, 2017).

In addition to poor relations between the protest leaders and key leaders in Imizamo Yethu and Hangberg, there is a further reason for this lack of participation by black community leaders of Hout Bay. In terms of the May 2002 concession agreement, Entilini formed a Community Trust with trustees from Imizamo Yethu, Hangberg, Masiphumelele, Westlake, and Red Hill. It was allocated 3 per cent of the concession company and 'as soon as the project is generating free cash flow to pay dividends, the trust will receive a regular income that will be managed by the community trustees' (Dreyer et al. 2005: 9). In addition, BEE spending during the operations phase of the project had to be at least 15 percent, excluding salaries. In 2005/6, the shareholders of the concession company provided R50,000 to the trust from their own social responsibility budgets (Yeld 2009). The Chapman's Peak website reports that:

> The Chapman's Peak Community Trust is a legal entity which has been established to undertake social responsibility projects or programmes in the Chapman's Peak area and to align all community projects in order to obtain a holistic overview of community development. The Trust, together with the Red Hill Community Forum, a charity organisation in the Chapman's Peak area, coordinates development programmes such as church organisations, youth programmes, the women's organisations and education programmes.

Indeed, both Imizamo Yethu leaders aligned to the ANC and Hangberg leaders in the PMF confirmed the existence of the Chapman's Peak Development Trust, and both specified that its role was awarding scholarships for students from the two poor, black settlements to study at university (Community leader 2 2017; PMF leader 2017). With a stake in Entilini, and direct participation by key community leaders in the Chapman's Peak Community Trust, it is not surprising that the leaders of Imizamo Yethu and Hangberg did not support the Toll Road protests with any real enthusiasm.

Notably, the Chapman's Peak Community Trust keeps a low profile in the public life of Hout Bay. Thus, other than the scholarships, the Trust has not succeeded in driving the promised 'social responsibility projects or programmes', and aligning 'community projects in order to obtain a holistic overview of community development' as suggested by the Chapman's Peak website. Not only is the Trust not on the exhaustive list of Hout Bay stakeholders compiled by the Hout Bay Partnership, it does not participate in the key development forums of Hout Bay including the Hout Bay partnership, the Hout Bay health forum, the Imizamo Yethu civil society forum, and the like. Indeed, the only presence we could find of the Chapman's Peak Community Trust in the public domain was on the Chapman's Peak website.

The construction of a mutually beneficial relationship between business and political leaders in the Entilini deal was not limited to community-based leaders, but also existed at the highest level. The clearest manifestation of this was the fact that Dr Lionel Louw, uncle to Greg Louw, a current leader of the Hangberg Peace and Mediation Forum, was right-hand man to ANC Premier Rasool as head of the Premier's Office from 2004 to 2008. Notably, at the same time he was a director of Marib holdings, which had a 10 per cent share in Entilini. A direct coincidence of interests is offset by the fact that the Entilini contract was awarded in 2002, and Lionel Louw became Head of the Premier's Office when the ANC came to power in the province from 2004 to 2009 (Gosling 2012b). While we do not know whether this intimate relationship between business and political leaders was the idea of the majority shareholder at the time, Concor, or key ANC politicians like Louw, the alignment of national business and provincial political interests helps explain the imposition of a project on Hout Bay that benefitted elite interests rather than local residents.

Ignoring public participation

In addition to the co-option of local political leaders, another key tactic evident in the founding phase of the Chapman's Peak Toll Road was the finessing of formal state processes, in particular the requirements for public participation. In 2002, opposition to the Toll Road by the HBRRA through the courts prompted the Provincial government to instigate a full Environmental Impact Assessment (EIA) to inform the plan for the plaza (Western Cape Government 2012). Emergent from this were a number of recommendations, including for a single toll booth to reduce changes to the environment, which was ignored until opposition grew in the late 2000s (Hart 2003).

During 2003, the first round of public participation took place, and included adverts in national and local newspapers, including the *Sentinel*, inviting Interested and Affected Parties (IAPs) to register and provide comment. Three public meetings were held at Fish Hoek (18 June, 30 July), Hout Bay (19 June, 31 July) and with the Chapman's Peak Community Forum (2 August 2003). Reflecting on her experiences of the public meetings, an environmental activist 2 (2015) commented:

> I went to all of the public meetings which were a complete farce because people objected but their views were just rolled over. The main objections given at the public meetings were the expense, the damage to the mountain and that now it would be a toll road because it cost so much. If you just fix it like any other so-called dangerous public road you would not need a toll. Nobody spoke for it at any of the meetings, but [government] just nodded and smiled and carried on doing what they wanted anyway, which is what I have seen is the typical way of doing public participation. I was bitterly disappointed in the public participation process.

A Residents' Association leader (HBRRA Leader 1 2017) recalls a few people from Hangberg supporting the toll plaza, but the rest of the residents at the meeting opposing it. Notably, in the face of significant mobilisation in 2012, Provincial MEC for Transport Robyn Carlisle, criticised protesters for opposing aspects of the toll booth that were made public through the public participation processes – a claim that Len Swimmer of the HBRRA contested (Gosling 2012a). This incident reflects a tendency in South African law to equate the process of public participation with legitimating decisions, even when the public substantively disagrees with proposals discussed in the participation process!.

Courts and the environment

Under South African law, the main mechanism available for public participation around infrastructure development is the Environmental Impact Assessment report. Hence, given its history of using the formal processes of government and the courts as its main medium of challenging government decisions, the HBRRA relied mostly on this tactic to challenge the imposition of the Chapman's Peak Toll Road (Lawyer 2017; HBRRA leader 1 2017). In 2003, the HBRRA advocated strongly for participation in the EIA process around the Chapman's Peak Toll Road. A site visit was conducted with the Chapman's Peak Drive Environmental Monitoring Committee on 10 July 2003, the EIA was lodged at public libraries and posted online, and registered interested and affected parties were invited to comment. An activist decided to take action using the public participation process (Environmental activist 2 2015):

> So I went to the library which is what you're meant to do, and I saw about 20 people had been to register their complaints, so I decided to work with them to do something. We created an A5 summary of what was happening and what people could do. We got free sponsorship to make 1,000 of these, and we dropped them off at schools and shops. We only had about three days to respond. We got 500 back from schools in IY HB and the Valley! So we took all of these to Tasneem Essop's office, MEC for Environment and Planning, and we said the people of Hout Bay do not want this.

Despite these objections, Chapman's Peak Drive was reopened to traffic as a toll road on 20 December 2003, after investing costs of R160 million (Western Cape Government 2012). Following its reopening, Chapman's Peak Drive was closed for 55 days in 2004 after heavy rains in winter (1/2 the annual rainfall). In October 2004, Entilini unilaterally decided to withdraw the R20 concession card for trips on the Hout Bay–Noordhoek route (Swimmer 2014). By 2005, the environmental impact assessments were complete, and the Provincial government issued a formal Record of Decision, as required by environmental law, in favour of constructing 'a toll plaza at the quarry site at Kooëlbaai, and a second toll plaza at the Noordhoek end of the pass' (Western Cape Government 2012: 8). In 2006, Murray & Roberts acquired

Concor and became the major shareholder in Entilini (Murray & Roberts 2012). At roughly the same time the Provincial government changed, with the ANC leading the Western Cape from 2004 to 2008.

The decision to construct two toll plazas was further challenged by a number of groups and individuals, led mainly by the HBRRA. Key to the Hout Bay Residents Association objection was the issue of conservation and, more specifically, the impact of the Toll Road and booth on the Table Mountain Nature Reserve that was declared a World Heritage Site in 2000 (Lawyer 2017). It was in the context of the struggle over the Chapman's Peak Toll Road that the HBRRA came to portray the City as favouring development projects over the environment, and ignoring public opposition in the process. In a letter to the *Cape Times* on 1 September 2014, Len Swimmer (2014) explicitly links what he sees as 'pro-development decision with insufficient regard for the protection of ... natural resources', 'short terminism' linked to elections, a growing and 'disturbing disregard for public opinion', including the 'discouragement of public participation', and a concomitant increase in 'the centralisation of decision making' powers.

This relationship between development and growing *dirigisme* at the expense of the environment has been made frequently on the HBRRA's and Greater Cape Town Civic Alliance's websites over the last 10 years. As demonstrated in Chapters 2 and 6, the appeal to conservation and environmentalism has also been used to limit the settlement of further poor black residents in Hout Bay, and the upgrading of facilities in Imizamo Yethu. In the words of one respondent (Environmental activist 1 2017):

> What I see is that ... wealthy developers can just go ahead and do it because they have access to funds to push plans through and to get it done, and on the other side, Imizamo Yethu is in dire straits and some of the real issues are not being looked at.

Thus the discourse of environmentalism (as opposed to ecologism) serves to integrate the more conservative views of an older generation of white residents against both new forms of development, and poor, black migration into Hout Bay. It is close to what Draper (2003: 57) characterises as a 'colonial preservationist mind-set seeking to alienate indigenous people from nature both intellectually and materially'. This conservatism is also evident in the preference for formal and legal challenge, in contrast with the forms of direct action taken by Bronwen Lankers-Byrne and her fellow ecological activists.

Resisting closure, resisting the toll booth

Thus, to return to the plan for two toll booths in the mid-2000s, while the legal objection made by the HBRRA was successful in securing further rounds of public participation, it made little difference to the original decision. In 2008, a second Record of Decision was issued by the Minister of the Environment, Marthinus van Schalkwyk, approving the two toll plazas once again (Western Cape Government 2012). The failure of the HBRRA and the use of formal channels only was precisely what inspired environmentalists to take a different approach in 2010 (Environmental activist 1 2017; Environmental activist 2 2015). What followed was the use of a range of innovative tactics from petitions and marches, to hunger strikes and obstruction that were able to generate enough media pressure to force meetings between protest leaders and both the Provincial Premier, Helen Zille, and the CEO of Murray & Roberts, Henry Laas. While this disruptive politics also failed to stop the Toll Road, it achieved much more when combined with formal participation in invited spaces, bringing about a number of new concessions.

In addition to a desire for different tactics, these environmentalists advocated a discourse of ecologism explicitly distinct from ideas of conservation that carry colonial-era contrasts between wilderness and indigenous people. Thus, where the HBRRA based its objection to the toll road on the conservation of the Table Mountain National Park, the environmentalists objected to the toll road on the basis of the exploitation of both the local environment and local people. The account of ecologism is one that affirms the sustainable living of people within the larger ecology, which for them necessitates social justice. Thus, recycling goes hand in glove with food gardens, and the sharing of common resources. Indeed, as outlined in Chapter 2, a key activist sees environmentalism as an issue that can unite Hout Bay as:

> I am sick and tired of wealthy people living in opulence, owning the majority of the land, literally 95% or even more percent of the land and the majority in our country are living in squalor; and it's continued for 14 years since I've lived in Hout Bay.
>
> (Environmental activist 2 2015)

Key to sparking the new, and more radical, round of resistance was the closure of Chapman's Peak Drive in June 2008 for major upgrades and repairs. The construction work took over a year and the road was eventually reopened on 9 October 2009. Two facts are critical here: first, in terms of the original concession, Entilini could unilaterally close the road without consulting Province; and second, there was a clause in the contract whereby Province would continue to pay on 'expected traffic volumes' when the road was closed. Notably, the projected traffic figures were calculated by Concor's consultants, Stewart and Scott Pty Ltd. They had projected, month by month, the number of vehicles they believed would use the toll over the next 30 years (Gosling 2012b). Thus, for the period that the road was closed, Entilini was paid R59 million in 'compensation' by Province. Further and, in retrospect, perhaps unsurprisingly, 'expected traffic volumes' have never been met on Chapman's Peak Toll Road. The project estimated daily usage growing to 4,000 vehicles a day. Yet, according to the Chapman's Peak website, the reported usage in 2014, which was the highest yet, was 835,756, or an average of 2,290 per day, just over 50 per cent of the estimate.[2]

In 2009, frustration with the unilateral closure of a public road, which many in Hout Bay also saw as an attempt to extract resources from the Province, led to direct action. As one participant explains:

> In 2009 things came to a head as the road was closed and they did not seem to be doing anything. There was a fence across the road, right at the top, and you couldn't even walk past. I mean bloody hell, it's a public road! Carte Blanche did a story and there was a big *opskop* (brawl), and after that I met with some people from Noordhoek, and we agreed that I would organise people to march from the Hout Bay side and they would organised people from the Noordhoek side. We did the big March and met the MEC for Finance in the middle to deliver a memorandum. People were angry, yoh, they were angry. His tyres got let down, but anyway! When we got to the fence, they asked me to tell the crowds to calm down because Entilini were going to open the gate. They didn't come with their keys, so somebody brought their bolt cutters and cut the lock. The gates open and it's like the Berlin wall! It was classic.
>
> (Environmental activist 2 2015)

These events happened against the background of a review of the concessionaire contract towards the end of 2008, based on an investigation ordered by ANC Premier Lynne Brown. This process surfaced both Entilini's frustration at the 5-year delay in constructing the plaza, and Province's 'frustration with repeated (and lengthy) road closures, which the public, under the existing contract, had to pay for' (Western Cape Government 2012: 7). With the change of government to the DA in 2009, the contract was renegotiated and by March 2011 'the so-called Third Amendment agreement was signed, which brought an end to unilateral closures of the pass [by Entilini] and the burden on the public purse of compensating the concessionaire during closures' (Western Cape Government 2012: 8). Notably, while Entilini agreed to repay the R59 million, they were given 20 years to do so. Remarkably, since 2011, Chapman's Peak Drive has remained open almost continually (Chapman's Peak website 2016).

The final phase of the toll project commenced with the building on the toll plaza in 2012. The plaza was controversial both because of its proposed size at 800m[2] (by 2012 reduced to 610m[2]) and cost of R52 million, of which Government was to contribute R25 million to be recouped from toll fees (Western Cape Government 2012). The Provincial government justified the plaza on the basis of the need for a world class facility designed to host 57 staff, including 'management, administration, secretarial support, technicians, supervisors, route patrollers, toll booth operators, maintenance staff and a cleaner'. The government document notes that the delays in building the toll centre meant that these staff had to work for 8 years in temporary facilities that consisted of six shipping containers stacked two stories high and four fibreglass toll booths, which created difficult working conditions (Western Cape Government 2012: 9):

> Toll booth operators in particular struggle with the eight hour shifts in the fibreglass boxes, where temperatures soar in summer and plunge in winter. Eight years of using port-a-loos while the machinery of due process grinds through the gears has also taken its toll, and the operator estimates as many as 200 employees have departed over the years, about 50% of staff year on year, with working conditions a major factor in many resignations.

These reasons did not wash with opponents of the Toll Road. While recognising that the Toll Road was probably now an inevitability, as was some form of plaza, given that this was in the original contract, they resolved to oppose what they saw as the unnecessary and wasteful version of a toll booth (Environmental activist 1 2017). Indeed, to this end protesters proposed a number of alternative toll booths, including one commissioned by Bronwen Lankers-Byrne that was estimated at R2 million. Given her experience in 2002 and especially during the march of 2009, Lankers-Byrne was resolved to do her utmost and organised a series of protests against the new toll plaza on Saturday 22 and Saturday 29 January 2012 (see Figure 8.3).

The first protest saw around 1,000 protesters from Hout Bay come out against the new toll station. Poster slogans included, '54 million! Can build houses for IY and Hangberg people', 'Stay Away from our Mountain', 'picnics before profits', 'Jou Ma se Toll', 'Stay Away from our Mountain', '10 of us in one shelter, can you build houses instead of playing with money', 'Hands off chappies robber Carlisle', 'Murray & Robbers', 'Zille: where's

Figure 8.3 Anti-toll booth protest 2012

the democracy?'. Although most protesters were from the valley there were also some from Imizamo Yethu and Hangberg (Bennett 2012), including the 2007 SANCO leader. As suggested in Chapter 6, however, he was leader of a faction of SANCO not in good standing with the ANC locally, and with limited capacity to mobilise in Imizamo Yethu. As noted above, the ANC leadership were included in the Chapman's Peak Community Trust and thus had a vested interest in the toll road construction proceeding.

After the protest, Lankers-Byrne reported engaging over email with Murray & Roberts, and getting the distinct impression they were open to dialogue over the toll booth, only to discover that on the Thursday following the protest they were on the mountain digging. 'I was angry that they were deliberately misleading me. So then I decided, well I'll go to a hunger strike to see if they'll listen, to force further negotiations' (Environmental activist 2 2015). Thus, on Sunday 5 February 2012, the hunger strike described in the opening vignette began.

During the hunger strike and obstruction, media coverage of the Chapman's Peak Toll Road reached unprecedented levels, and supporters secured over 10,000 signatures in a petition calling for the cessation of the building of the toll booth (Bennett 2012). With the removal of hunger strikers from the site, building continued, but the protest did not end. On 2 March 2012, Entilini began felling a stand of gum trees on SANParks land and a group of 12 protesters tried to halt the felling until the police were called. In response, the HBRRA initiated an interdict to halt all construction as it was a World Heritage Site (Bennett 2012). According to Len Swimmer, the heart of the case was that usage of a global World Heritage Site should be for conservation, not for profit. More specifically, the objection was that the Records of Decision in 2005 and 2008 were made on old information from previous EIAs. Prof Merle Sawman of UCT wrote a letter to Helen Zille expressing these concerns during this period.

In March 2012, activists used the Cape Argus cycle tour and the Two Oceans marathon to protest the toll booth, as both events pass along Chapman's Peak Drive. Lankers-Byrne recalled how on the morning of the Argus cycle tour she crashed a media interview that Helen Zille was doing about the Argus to demand that she meet with her.

> She agreed and I had half an hour between 7:15 and 7:45 in the morning. She argued that it would cost R140 million to cancel the contract at that point, and it was too costly. I do think her hands were pretty tied by the contract. My thought was perhaps we could get them both to agree to review.
>
> (Environmental activist 2 2015)

Thus, just before the Two Oceans marathon, Lankers-Byrne went up to the head office of Murray & Roberts with her 96-year-old father and blocked the entrance to Murray & Roberts at Bedford View. She noted, there were three of us, but my friend had to take the pictures as none of the press had arrived then! The CEO, Henry Laas, invited us into a meeting with two of his Directors in the building. The signs were all put inside the room while we met! Nothing really came of the meeting, they insisted they were bound by the contract.

(Environmental activist 2 2015)

These two engagements with Zille and Laas reveal quite profoundly the loss of popular sovereignty involved in market governance. Through the device of any form of contract, in this case a Public-Private Partnership, service providers gain the coercive backing of the state without the accountability mechanisms that usually apply to elected officials. Thus, even the Premier of the Western Cape was unable to revoke unilaterally a decision on a government-sponsored project that was both unwanted by the residents it most affected and was costing the government money. Notably this was not just a problem for Zille, but also her predecessors, and it was not just Concor who benefitted from this, but also the subsequent investors who bought it out. Lastly, it was only popular resistance that brought costs for both business and political elites that enabled the renegotiation of the contract in the 'Third amendment agreement' of 2011.

After the meeting at Murray & Roberts, resistance to the toll road petered out. There was another protest at the High Court in Wynberg on 28 May 2012 when the state attempted to prosecute Fiona Hinds for spray painting 'Murray & Robbers' at the construction site, but the case was dismissed by the Court. In June 2012, the Residents Association court case failed on the basis that the likely damage to the World Heritage Site was not significant (Residents Association of Hout Bay and Others v Entilini and Others, 2012). The only glimmer of justice in this respect was that, in 2013, Murray & Roberts paid a R309 million fine in terms of the Competition Commission's fast-track programme in connection with price collusion around the 2010 World Cup, leading the HBRRA to appeal to the Competition Commission to investigate the Chapman's Peak concession (Bernardo 2015).

Conclusion

The decision to toll Chapman's Peak Drive through a PPP between Provincial government and national construction companies is an exemplary case of governance rather than government. Province co-governs transport in a formal partnership with other social actors, in this case with a national business. It is also a profound illustration of how powerful actors external to the City of Cape Town can shape the ways in which people must live in the city. Lastly, it is an exemplar of market governance where a public resource is privatised for reasons of financial viability. Notably, these decisions were not popular, and consistently opposed by many residents of Hout Bay, if not necessarily most of those affected by it.

The ensuing struggle between Entilini, the Provincial government and residents sheds important light on the nature of market governance, the disconnect with local democracy, and the resultant bifurcation of public participation into co-option through representation or disruption through mobilisation, as well as the variegated nature of green politics in Hout Bay. In respect of market governance, the Chapman's Peak case shows how private actors are enabled by law to retain the coercive backing of the state, but without the requirement of popular accountability. Thus, the initial attempt in 2000 to make Chapman's Peak Drive safe was changed at the prompting of big business, and led to the passing of new PPP legislation by Provincial government and the award of a contract to the same business with uncommon speed. A few years later, when the DA came to power in the province, it was clear that both the residents of Hout Bay did not want the Toll Road, and the Provincial government were economic losers in the original deal. Despite this, the new Premier was unable to change things unilaterally, because under market governance, sovereignty resides in contract law, not with the legislature, and certainly not with the people.

In addition to offering insights into market governance, the tolling of Chapman's Peak illustrates the limitations of invited spaces of public participation to influence decision-making. Partly this is because local democratic institutions have no purchase over Provincial government, and partly because processes of public participation on development projects are not binding. In addition, project managers co-opted important leaders from Imizamo Yethu and Hangberg onto the Chapman's Peak Development Trust, creating a vested interest in the Toll Road. Frustrated by the unresponsiveness of formal institutions, activists turned to protest and disruption, illustrating how, when democratic engagement is weak, public participation tends to either representation through co-option, or mobilisation to disrupt.

Lastly, in terms of popular politics, the chapter affirms the importance of green issues to the conceptions of being and belonging of white and wealthy residents of Hout Bay. Ideologically this stretches from accounts of conservation through environmentalism to ecologism, with different notions of how nature and society should relate. For many, the natural beauty of Hout Bay should be preserved, and the environment protected from human development that pollutes the river, destroys the natural bush and so on. For others, it is about finding new ways for residents and nature to coexist sustainably. Important to the Hout Bay story, however, is how key organisations in the Valley have invoked conserving the environment to justify opposing both developments like the Toll Road and the in-migration of poor, black residents. By linking environmental threats to identities, this discourse reinforces racial conceptions of belonging in Hout Bay identified in Chapter 6. While it is not the only form of green politics in Hout Bay, and the ecological view of being in Hout Bay affirms inclusive conceptions of social justice, the latter is very much a minority view.

Notes

1 Western Cape MEC for Transport Robin Carlisle took the *Cape Times* to the Press Council (2012) on a range of issues, including the estimated cost of R360 million which he insists was R160 million. Notably, the figure of R360 million is taken from the conference paper of Dreyer et al. (2005), the engineers of Entilini.
2 See www.chapmanspeakdrive.co.za/traffic-stats.html.

References

Bennett, A. (2012). 'Site seeing'. Documentary on Chapman's Peak protest. 27 April. www.youtube.com/watch?v=WmZuyCI0WW8&feature=youtu.be. Accessed 17 January 2016.

Bernardo, C. (2015). Murray&Roberts fined R64 million. *Independent Online*, www.iol.co.za/business/companies/murrayroberts-fined-r64-million-1958709. Accessed 21 November 2016.

Cape Metropolitan Council v Noel Raymond Graham (157/99) [2000] ZASCA 17 (14 November 2000). www.justice.gov.za/sca/judgments%5Csca_2000/1999_157.pdf. Accessed 17 January 2016.

Chapman's Peak website. (2016). www.chapmanspeakdrive.co.za/. Accessed 15 October 2017.

Draper, M. (2003). 'In quest of African Wilderness', USDA Forest Service Proceedings RMRS-P-27. 57 https://www.fs.fed.us/rm/pubs/rmrs_p027/rmrs_p027_057_062.pdf. Accessed 10 July 2018.

Dreyer, W., Breytenbach, K., Watters, M., Van Oudenhove, W., & Parring, P. (2005). Innovative PPP saves Chapman's Peak: PPP brings together the public and private sectors for rehabilitation of the famous road. Proceedings of the Southern African Transport Conference (SATC), Pretoria, 11–15 July. http://repository.up.ac.za/bitstream/handle/2263/6594/051.pdf?sequence=1. Accessed on 17 January 2016.

Eliot, H. (2013). Letter: truth first casualty in 'Chapman's Peak war'. *BDlive*, 27 September 2013. www.bdlive.co.za/opinion/letters/2013/09/27/lettertruth-first-casualty-in-chapmans-peak-war. Accessed on 17 January 2016.

Gosling, M. (2012a). Carlisle's U-Turn on Chapman's Peak. *IOL Online*, 12 January. www.iol.co.za/capetimes/carlisles-u-turn-on-chapmans-peak-1211703. Accessed 6 November 2016.

Gosling, M. (2012b). Taking a drive through the Chappies saga. *IOL Online*, 30 January. www.iol.co.za/news/south-africa/western-cape/taking-a-drive-through-the-chappies-saga-1.1223531?ot=inmsa.ArticlePrintPageLayout.ot. Accessed 17 January 2016.

Hart, T. (2003). Chapman's Peak Drive – heritage impact assessment of proposed locations of toll plazas. Environmental Impact Assessment prepared for Ninhand Shamd consulting Services, September. www.sahra.org.za/sahris/sites/default/files/heritagereports/AIA_Chapmans_Peak_Drive_Tolls_Hart_TJ_Sep03.pdf. Accessed 17 January 2016.

Murray & Roberts. (2012). Murray & Roberts sets the record straight on Chapman's Peak. *Murray & Roberts News*, http://web.murrob.com/mobi/news_Murray_Roberts_sets_the_record_straight.html. Accessed 17 January 2016.

Press Council. (2012). Ministry of Transport and Public Works (Western Cape) v Cape Times. Press Council Ruling, 24 July. www.presscouncil.org.za/Ruling/ View/ministry-of-transport-and-public-works-western-cape-vs-cape-times-2340. Accessed 10 January 2016.

Residents Association of Hout Bay and Others v Entilini and Others (7648/12) [2012] ZAWCHC 23 (6 June 2012).

Swimmer, L. (2012). Len's Lines. *Hout & About*, Hout Bay Residents' and Ratepayers' Association, www.houtbay.org.za/RAHB_Newsletters.html. Accessed on 17 January 2016.

Swimmer, L. (2014). Centralised decisions destroying the City's natural resources. *Cape Times*, 1 September 2014. http://gctca.org.za/centralised-decisions-destroying-the-citys-natural-resources/. Accessed on 17 January 2016.

Western Cape Government. (2012). Chapman's Peak Drive: Operations Centre and Toll Plaza. Ministry of Transport and Public Works. www.westerncape.gov.za/ other/2012/3/chapmans_peak_march_2012_v2.pdf. Accessed 17 January 2016.

Yeld, J. (2009). Probe into Chappies Deal. *IOL Online*, 10 February. www. iol.co.za/news/south-africa/probe-into-chappies-deal-1.434008?ot=inmsa. ArticlePrintPageLayout.ot. Accessed 17 January 2016.

Interviews

1. Community leader 2. (2015). Interviewed by Laurence Piper. 20 May 2015.
2. Community leader 2. (2017). Interviewed by Laurence Piper. 24 August 2017.
3. Environmental activist 1. (2017). Interviewed by Laurence Piper. 24 August 2017.
4. Environmental activist 2. (2014). Interviewed by Laurence Piper. 17 February 2014.
5. Environmental activist 2. (2015). Interviewed by Laurence Piper. 30 November 2015.
6. HBRRA leader 1. (2017). Interviewed telephonically by Laurence Piper. 23 August 2017.
7. Lawyer. (2017). Lawyers representing the HBRRA. Interviewed telephonically by Laurence Piper. 24 August 2017.
8. PMF leader. (2017). Interviewed by Laurence Piper, 24 August 2017.
9. Zille, H. (2017). Premier of the Western Cape. Interviewed by Laurence Piper. 26 September 2017.

9   Guarding the bay

Securing safety beyond the police

On a hot and dusty Tuesday afternoon we are standing at the edge of Imizamo Yethu, where the township meets the police station, talking to a local community leader. The days leading up to this conversation have been rife with violence and conflict in Imizamo Yethu. Over the previous few months, the settlement has seen the rise of gangs and attacks on residents.

On 8 July 2015, 27-year-old Nchikala Ngoy was fatally stabbed eight times in the back and head while walking home from the library. His cell phone was stolen (Genever 2015). One month later a highly regarded 16-year-old school student and his 22-year-old companion, who worked at the local version of Starbucks, were murdered as they walked through the settlement to buy electricity (Legg & Kalipa 2015). Both events were linked to the drug gangs, the amaXaba and the Bad Boys Company.

The community leader tells us how, in response to these murders a vigilante group run by the community formed. 'The community decided to stop the gangs ourselves'. We ask our informant what happened. He explains that the vigilante patrollers marched through the settlement and 'were confronted' by the leaders of the two gangs in Imizamo Yethu: 'The gang leaders were stoned to death. We met with the gang members afterwards to make peace. Many gang members did not know they had relatives on the other side' (Community leader 2 2015).

The surreal nature of this conversation highlights the extent to which violence is commonplace in a township such as Imizamo Yethu, and indeed crime and insecurity are two of the largest concerns facing all residents of Hout Bay (Piper & Wheeler 2016). This chapter will look at the rise of crime in Hout Bay and the impact this has had on how residents inhabit and experience the spatial and identity boundaries of Hout Bay. Specifically, it will look at what forms of governance are employed to address crime across all three areas of Hout Bay.

We demonstrate that the forms of governance that provide safety and security in Hout Bay are largely independent of the state, managed through private or informal control. This mirrors, to some extent, existing research on network and nodal security governance (Johnston & Shearing 2003; Boutellier & van Steden 2011). However, moving beyond the conceptual limitations of nodal governance, we demonstrate that, while the state has limited power over crime governance, actors in this field do work together with state agencies; indeed network governance or co-governance is a key feature of the crime fighting space in Hout Bay. In conclusion the chapter shows that one of the outcomes of private and informal governance over safety and security is the confirmation of neo-apartheid residential segregation in governance terms too.

Crime and violence in South Africa and Hout Bay

It was widely believed that violence (often politically related to defending and overthrowing apartheid) in South Africa would start to wane after the democratic transition in 1994. This has not been the case. South Africa has maintained its violent culture; now, however, it manifests in acts of crime rather than political rebellion. South Africa has a murder rate of 33 per 100,000, which is five times the global average (Kriegler & Shaw 2016). The Western Cape has a more pronounced problem with a murder rate in Cape Town of 65 deaths per 100,000 (Hosken 2016). Cape Town certainly has the highest murder rate in the country and is also ranked as the ninth most violent city in the world (BusinessTech 2016). While the high murder rate is largely attributed to gang warfare in specific parts of the city, not including Hout Bay, it nonetheless highlights the enduring nature of crime and violence in South Africa.

In keeping with the national trend, Hout Bay also experiences high crime levels. Although levels of crime have been relatively stable or have even dropped, crime is nonetheless at worrying levels. The graph below (see Figure 9.1), derived from police station data (SAPS n.d.), provides some indication of the extent of crime of Hout Bay. However, using statistics to understand crime (aside from that of murder) can be problematic (Kriegler & Shaw

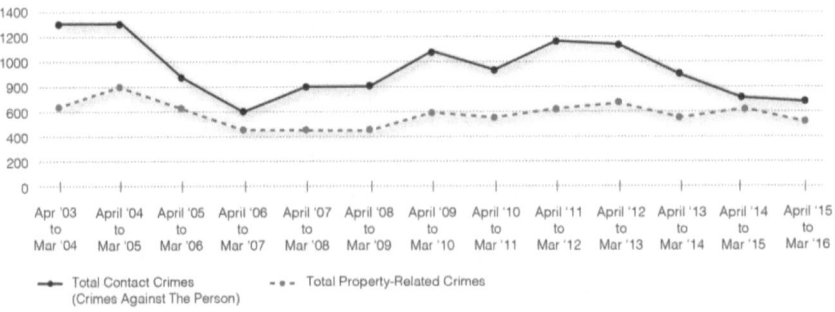

Figure 9.1 Levels of reported crime in Hout Bay, 2003–2016

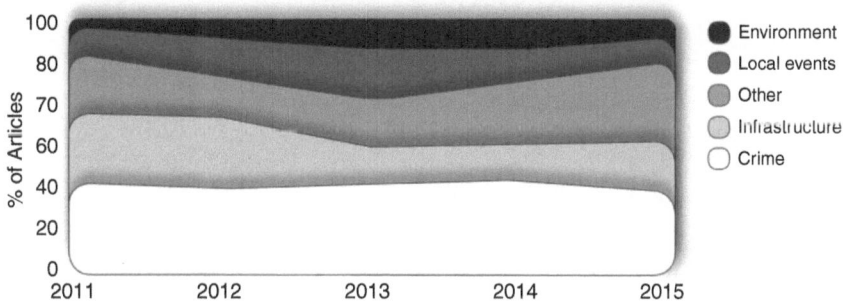

Figure 9.2 Sentinel newspaper major issue types, 2011–2015

2016). There is often under-reporting of crime such as rape and domestic violence, or of crimes that are quickly resolved where the victim does not open a case at the police station (SSP patroller 2016). As will be discussed below, crime in Hangberg and Imizamo Yethu is significantly under-reported.

Certainly, crime is an issue that residents in Hout Bay feel is central to their daily lives. A survey we conducted of the local weekly newspaper, the *Sentinel*, shows that over the last 5 years, crime has been the most significant issue that the paper reports on (see Figure 9.2). In 2015, 35 per cent of issues reported on in the *Sentinel* newspaper related to crime and security.

Governing crime

Bureaucratic governance: the South African Police Service

The primary way in which the state manages crime is through policing. Hout Bay has one police station under the jurisdiction of the South African Police Service (SAPS). The SAPS was established in the Constitution, and national legislation frames its powers and functions. Political responsibility for policing is vested at the national level in the Cabinet, although national policing policy is determined after consulting the provincial governments. Operationally, SAPS is run nationally, with the President appointing a National Commissioner of the police service. The Commissioner appoints a provincial commissioner for each province, who is responsible for provincial policing matters. Overall, however, the control of police stations in South Africa remains a national mandate, and local government has no authority over their functionality. SAPS is a government department that works according to Weberian bureaucratic principles, hence we use the term 'bureaucratic governance' to describe the way in which it make decisions over issues of safety and security.

The responsibility for allocating resources to a police station thus lies with national government. As proponents of the concept of 'anchored pluralism'

Figure 9.3 SAPS station in Hout Bay

(in relation to security governance) argue, the state has the capability to assure an 'accurate (re)allocation of collective resources by enabling agents and agencies ... to alleviate feelings of ontological security and anxiety' (Boutellier & Van Steden 2011: 467). The allocation of police resources is a significant point of contention in the Western Cape, and indeed in Hout Bay. According to the Western Cape Ministry of Community Safety, there is a 'disastrous police-to-population ratio' in many of the areas, with high rates of crime in Cape Town (Plato 2013: n.p.). The national average police-to-population ratio is in the range of one police officer for every 303 citizens; however, in an area such as Mitchell's Plain, with 66 murders in 2011/2012 there was a ratio of 3,240 citizens per police officer (Plato 2013). While not as bad, Hout Bay has similar problems, with respondents citing the poor police to citizen ratio and chronic understaffing of the police station as key concerns (SSP patroller 2016; Captain 2016; CPF Exec 2016).

According to the DA's Western Cape spokesperson on community safety, Hout Bay has a ratio of 506 citizens per police officer (Raba 2016). The Hout Bay station has approximately 20 administrative staff and 70 uniformed officers. Fifteen of these are detectives, while others are allocated to functions such as managing firearms and liquor licences, police exhibits, human resource functions and crime prevention planning. Of a shift, there are only six to seven active officers on duty. Two officers will staff the police station desk while four or five will be on patrol. There are only two SAPS vans in Hout Bay

(Captain 2016). Active SAPS officers may also be deployed to other parts of Cape Town if there is a need for additional policing, for example, to manage protests. This means that, at times, there may only be two officers actively patrolling Hout Bay. If a van is out of commission this also poses a serious problem to visible policing (CPF Exec 2016).

SAPS is not the only state entity dealing with safety and security in Hout Bay. South African National Parks (SANParks) are responsible for policing the land they manage on the mountains. They have approximately 30 trained officers, with a base in Hout Bay. However, the area they patrol is vast. They mostly respond to threats to safety on the mountainside (Captain 2016), but also work with the DAFF to provide law enforcement for Marine Protected Areas (SANParks n.d.). Allied to this is the Hout Bay Harbour Master who, in theory, must record arriving vessels in the harbour. This, however, seldom happens, and the harbour is poorly controlled (Captain 2016; Poachers 2016). In terms of the City, Hout Bay has six City law enforcement officers based in the ward to enforce City by-laws. These officers work office hours and, particularly in summer, spend time monitoring the beachfront. Metro police are also responsible for managing law enforcement of the City council houses in Hangberg (Captain 2016). Given limited resources, it is not surprising that state agencies struggle to provide adequate safety and security in Hout Bay. It is thus essential that the state look to other models of policing.

Developmental governance: community policing

The legacy of apartheid policing, where law enforcement was responsible for upholding the authoritarianism of the state, has undermined the relationship between communities and the police. With a predatory police force, residents, particularly in townships, have increasingly turned to community structures to enforce safety and resolve conflicts locally (Tefre 2010). Thus, post-apartheid, the state is faced with rising crime levels, a mistrusted SAPS, and relatively strong community policing structures. One strategy the state has adopted to tackle this trifecta is the idea of community policing, a form of governance as the co-production of services which aims to improve police services and police legitimacy through proactive reliance on community resources, which change crime-causing conditions (Rosenbaum & Lurigio 1994). Community policing forms part of a nationally mandated 'grand project' to improve safety and security. In the co-production between state and community and in the linking of policing to national objectives of the developmental state, framed in collective and 'biopolitical' terms of populations to be managed (Chatterjee 2004), we have a basis for the idea of developmental governance.

While community policing exhibits characteristics of developmental governance, it also relates to, and stretches, debates on the 'nodal' theory of security governance. As we discuss in the Introduction, nodal governance essentially argues that there is a proliferation of nodes (or organisational sides of security) that may be public, private, or hybrid in nature. Nodes may be a point on a network, may have a territorial base (such as a shopping mall), or may simply be part of a virtual community (such as a Facebook site) (Boutellier & Van Steden 2011). Nodal theory has, however, been critiqued for minimising the centrality and authority of the state in security governance. In response, proponents of anchored pluralism (Loader & Walker 2006) argue that the state plays both an authoritative and symbolic role in relation to security. This is enacted by the state, for example, when it supports and initiates community policing.

The central element of community policing in South Africa is the Community Policing Forum (CPF). Hout Bay has had an active CPF for two decades (Captain 2016). CPFs were made mandatory in the 1993 Interim Constitution, which also gave them important powers of accountability over SAPS police stations. The 1997, policy on community policing stressed that CPFs should be involved in improving service delivery and facilitating partnerships for problem-solving (Pelser, Schnetler, & Louw 2002). CPFs are guided by a National Community Policing Desk, which falls under the National Crime Prevention and Response Unit (Pelser 1999).[1]

Following the national guidelines, the CPF acts as a link between the police station and the wider Hout Bay community. The main CPF in Hout Bay has set up three sector sub-forums corresponding to the suburbs of Imizamo Yethu, Hangberg and the Valley. The CPF and its sub-forums are not involved in direct policing or crime control activities: 'the CPF is just there to make sure there's enough crime prevention programmes in place and to watch how the police are doing their job' (CPF member 2016). The CPF holds regular meetings, has an active Facebook page, and the Chairperson conducts frequent oversight visits to the police station (CPF Exec 2016). There is significant overlap between elected members of the CPF and organisers of various neighbourhood watch groups. The Vice-Chair of the CPF sub-forum in Hangberg is also the sector leader for the Hangberg Neighbourhood Watch. In Imizamo Yethu, leaders of the 'patrollers' are in the executive of the CPF, and in the Valley the Chair of the CPF is a sector leader for the Hout Bay Neighbourhood Watch (HBNW) (HBNW Exec 2016). By all accounts, the Hout Bay CPF has a strong membership and a good working relationship with all other state and non-state actors involved in safety and security in Hout Bay. Certainly, in an interview with a Hout Bay SAPS Captain (2016), he stated that the Hout Bay CPF is seen as a role model in the Province.

The CPF is an important form of participatory democracy, but also is a form of co-governance with the potential to offer more control over policing to local residents. The Hout Bay CPF's role is largely one of co-ordination and communication between various crime fighting entities. Further, CPFs face numerous challenges, including a lack of basic resources and a lack of systematic and practical support from government, with community policing seen as an 'add-on' function to other police responsibilities (Mottiar & White 2003; Pelser 1999). Certainly, the Hout Bay CPF works on a small budget with volunteerism as its backbone. Indeed, the voluntary nature of citizen participation and the fear of retaliation by perpetrators may affect community involvement negatively (Mottiar & White 2003). This largely advisory role of the CPF is inadequate to tackle crime and insecurity in Hout Bay, and thus residents feel a need for more proactive forms of governance.

Network governance: the Neighbourhood Watch

It is clear that the state's ability to deal with crime is inadequate and residents therefore turn to alternative forms of governance to deal with crime and the provision of security. Indeed, the concept of community policing in South Africa extends to the idea that residents, as well as the private sector, are supposed to supplement the police in a context where police resources are seen as insufficient to fulfil the task (Bénit-Gbaffou 2008). This resonates with concepts of network and nodal governance. The key form through which residents organise independently to confront the issue of the crime is the neighbourhood watch.

A neighbourhood watch can range from an informal discussion group on crime to the active patrols and enforcement we see in Imizamo Yethu. In Hout Bay, the large and active Hout Bay Neighbourhood Watch (HBNW) works across all areas of the ward with 27 sectors grouped into eight blocks, including blocks for Imizamo Yethu and Hangberg. The HBNW is formally constituted as a sub-forum under the CPF, and is an active member of the organisation (HBNW Exec 2016; HBNW website). HBNW initially started in 2005 as a response to both rising crime levels in Hout Bay, and the collapse of the relationship between the CPF and the Hout Bay SAPS Superintendent.

In terms of the former, crime in the Valley had been on the increase since 2000. As one interviewee noted, 'in 2000 when I came here crime was very bad, people getting burgled left right and centre' (SSP director 2016). In 2003, the crime rate increased markedly across all categories, with property-related crime (i.e., housebreaking and theft out of vehicles) a particular concern. The catalyst for the formation of the HBNW was, however, the murder of a tourist in a Hout Bay guesthouse during a robbery. This event triggered the fears of those living in the Valley in particular, with the launch of HBNW attended by nearly 500 people. A year after the launch, HBNW had 1500 members, with over 100 involved in active patrols at night (SSP director 2016; Tefre 2010; HBNW website).

In its earlier formation, the HBNW was mostly a vehicle to protect the safety and security concerns of the wealthier residents of the Valley. The Watch did not initially involve Hangberg and Imizamo Yethu. This has changed significantly, and in 2016 the Watch had very active patrollers in both Imizamo Yethu and, to a lesser extent, Hangberg (CPF Exec 2016; Patrollers 2016). HBNW now has over 3,000 members (HBNW exec 2016). The core of HBNW is WatchCon, a central monitoring station that members of HBNW can call into to report suspicious activities and active crime.

It is important to note that HBNW is relatively well-resourced. Alongside a significant monthly donation from the private security company, ADT, members are also encouraged to pay a monthly fee for WatchCon. Patrollers in Hangberg, however, feel that the resources are largely spent in the Valley, with little relative support for Hangberg (Pastor 1 2015). Patrollers in Imizamo Yethu also explained that they are under-resourced and would benefit from resources such as reflective jackets, boots, and torches (Patrollers 2016). According to the HBNW Executive member (2016), however, resources are fairly allocated and there is no tension around how money is spent. The decent resources of the HBNW are in stark contrast to the limited resources of SAPS. They highlight, in material terms, the emphasis that residents place on civil society-led security governance in the context of an inadequate state system, and the importance of having a selection of 'nodes' that can network to improve safety and security.

The HBNW claims that its members are there simply to be the 'eyes and ears' of SAPS and Security Service Providers (SSPs) in Hout Bay. 'You are encouraged to go about your day and your Patrols, calling-in any suspicious vehicles or people behaving in a suspicious manner which you may observe, and in so doing allowing the relevant Authorities to investigate further' (HBNW welcome letter 2016). In practice, the HBNW activities differ greatly between the three areas of Hout Bay. In the Valley, most activity by members is largely 'eyes and ears'. However, there are more active members of the Watch called 'first responders' who 'want to get involved physically and can rush to a situation' (HBNW Exec 2016). These members are actively linked to WatchCon and can be first on site in their immediate neighbourhood. They may conduct a citizen's arrest, or detain a suspect where they feel it is warranted (SSP director 2016; HBNW Exec 2016). This active form of private security governance at times blurs the line between community policing and vigilantism, and can be seen as a type of informal governance.

HBNW is an important non-state 'nodal' or 'network' actor in the implementation of security governance in Hout Bay. Certainly, the formation of HBNW did lead to a drop in crime statistics in Hout Bay, as Figure 9.4 illustrates. The Watch went through a less effective phase after 2008 but has recently picked up its activity. The HBNW plays an important role in reducing crime in Hout Bay, certainly across the Valley and Imizamo Yethu. Local crime prevention organisations (CPF and HBNW) and the SAPS are adamant that crime in Hout Bay has dropped in the last 18 months in most parts of the ward. They also believe that crime is lower in Imizamo Yethu than other townships in Cape Town (SSP patroller 2016; CPF Exec 2016; Patrollers 2016).

Market governance: private security service providers

While CPFs and neighbourhood watches are important actors in the field of safety and security, so too is the private security industry. It is a major actor in

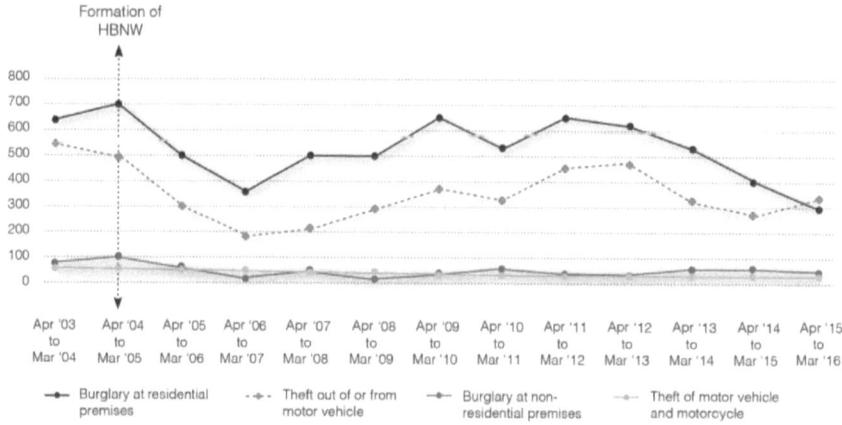

Figure 9.4 Contact Crime in Hout Bay, 2003–2016

terms of security governance in South Africa, in economic terms, accounting for 2 per cent of the national GDP (Diphoorn & Kyed 2016). As early as 1999, it was estimated that there were more than four private security guards for every uniformed member of the SAPS engaged in visible policing work. South Africans are therefore far more likely to encounter private security guards than police officers (Irish 1999). The private security industry has grown year on year due to increased insecurity and fear of crime coupled with the perception of a weak police service. However, deliberate neo-liberal policies implemented by the state that support outsourcing, along with a national crime strategy that encourages community and partnership policing, have also led to the growth of the industry (Diphoorn & Kyed 2016). Indeed, for most citizens in suburban Africa, 'private security companies (PSCs) and civilian self-help or community policing groups constitute the most significant everyday providers of security. This challenges the idea of the state as the only legitimate pro-vider of security and the welfare-state notion of security as a free public good' (Diphoorn & Kyed 2016: 719).

The two main SSPs working in Hout Bay are Deep Blue and ADT (a sub-sidiary of the American company Tyco International).[2] There is also a smaller community-funded organisation that focuses on patrolling, set up by previous members of the HBNW, called Community Crime Prevention (CCP). Deep Blue was started in 2003 when its founder moved to Hout Bay and saw a gap in the market for a community-based SSP (SSP director 2016). Indeed, being perceived as 'part of the community' is an effective marketing tool and SSPs in South Africa often articulate notions of community to create and attract collective clients (Diphoorn & Kyed 2016). While some forms of non-state, or nodal security governance such as CPFs and neighbourhood watches are not expressly linked to market governance, SSPs are. In their relationship with

SSPs, residents are acting as individual consumers. Safety and security are now bought and sold in a competitive marketplace. The public, as consumers, have the right to 'shop around' for the best quality service (Abrahamsen & Williams 2007: 134), and certainly in Hout Bay there is fierce competition between the two main service providers. This competition is in fact encouraged by the CPF as, 'The more they compete with each other the better and cheaper services we get in the community' (CPF Exec 2016).

Although SSPs are market actors they do still have a good working relationship with the SAPS and other non-state security actors, such as HBNW and CCP. For example, the SSPs are invited to weekly meetings at the police station to discuss crime prevention and co-ordination efforts (SSP director 2016). This echoes theories of urban co-governance discussed in the Introduction. Further, it appears to work. There is little doubt that the increase in private governance over safety and security in Hout Bay has led to a reduction in crime. For the vast majority of residents this is the most important outcome of the privatisation of security and is thus, from this perspective, a success story. However, formal and legal privatisation of security is only one element of the success story. Informal forms of governance play a significant role, too.

Informal governance: from vigilantism to crime lords

Neighbourhood watches and private companies, as individual nodes in the network of security governance, may not be state agencies, but they are recognised in law as formal partners in security governance. They adhere, largely, to the legal requirements of the state. However, a growing number of studies of local non-state actors (including vigilante groups, gangs and abrasive citizen-led policing organisations) have shown how 'different entanglements between civilian policing actors and the state police not only reconfigure sovereignty, but also blur the state/non-state and formal/informal divides' (Diphoorn & Kyed 2016: 716). The uneven capacity of the state, the inability of SAPS to control crime, and the visible successes of civil society and private companies are several reasons why we see, in all three areas of Hout Bay, the embracing of informal, and (or) illegal, tactics to fight crime. Nodal theorists note that it is possible for 'deviant nodes' to form 'dark networks' such as mafia syndicates involved in criminal activities (Raab & Milward 2003). This chapter, however, takes a less pejorative view of the informal governance of security. It looks at how informality manifests in three distinct ways: through spontaneous and sporadic vigilantism in Imizamo Yethu; the internal policing of crime in Hangberg; and through the illegal construction of road closures and booms alongside zealous neighbourhood watch activity in the Valley.

Imizamo Yethu

Imizamo Yethu, similar to most townships in South Africa, has high levels of crime and insecurity. The levels of contact crime are lower in Imizamo Yethu than in most other townships in Cape Town. Evidence for this is the average murder rate in Hout Bay of 13 people per year (the majority of these murders in Imizamo Yethu), equating to 26 per 100,000, lower than the national average of 33 per 100,000 (CPF Exec 2016). However, this is still high by global standards. Further, it is important to note that respondents participating in workshops in Imizamo Yethu talked about intimate crimes they had not reported to the police or even shared with their families. Thus, many crimes, principally those related to sexual violence, mugging, and robbery are significantly under-reported (Piper & Wheeler 2016; CPF Exec 2016). Over the past decade, there has also been a rise in drug-related crime with many respondents reporting, in 2012, the emergence of drug gangs for the first time in Imizamo Yethu's history. Certainly, residents of Imizamo Yethu fear crime, especially at night, with one woman explaining: 'We live in fear of someone kicking down the shack door and raping us' (Piper & Wheeler 2016). A Hout Bay SAPS Captain (2016) confirmed that crime in Imizamo Yethu was historically 'out of control' and that it was consistently the 'crime hotspot' for Hout Bay.

Today, however, that story has changed, at least to a degree. As described in the opening of this chapter, a series of senseless murders and ongoing muggings and rapes have triggered a strong community reaction. According to the current Imizamo Yethu patrollers (Patrollers 2016), it felt like there was 'a murder almost every week'. By August 2015, there was an ongoing battle between two rival gangs in Imizamo Yethu, the amaXaba and the Bad Boys Company. They were led by relatively young men and many of the members were senior school age children, and even younger. They were 'attacking each other, which would start at parties, and many had a serious drug problem' (Patrollers 2016). As described in the opening of this chapter, one night, in August 2015, a large vigilante group, with support from local community leaders, confronted and killed the leaders of both gangs (Community leader 2 2015). In the days that followed leaders from SANCO met with the remaining gang members to end the conflict and demobilise the gangs (Piper & Wheeler 2016).

Between August 2015 and early 2017, a regular, usually nightly, patrol operated in Imizamo Yethu. It started with hundreds of men from the community: 'so many they could not fit in the street' (Patrollers 2016). In time, this dropped to a group of between 50 and 70 patrollers who, in smaller shifts, walk the streets of Imizamo Yethu. There was a nightly curfew, they ensured that shebeens are closed by midnight, and they responded immediately to any suspicious activity or potential crime. The patrollers also reported to the SAPS that they were on duty every evening, and each had a bright bib and an identity card (Captain 2016). This was not the first time a community watch group had operated in Imizamo Yethu, with iterations from 2000, 2003 and 2007 functioning at different levels of success and sustainability (Patrollers 2016; Tefre 2010; Piper & Wheeler 2016).

When talking to actors involved in crime prevention in Hout Bay, it is clear that they do not see the patrollers of Imizamo Yethu as vigilantes, but rather as active and effective members of the CPF and HBNW (SSP patroller 2016; CPF Exec 2016; HBNW Exec 2016). Certainly, they have reduced crime in Imizamo Yethu significantly. A senior member of the CPF (2016) notes,

a lot of these youngsters were standing on the street corners ... and stealing handbags from local people in Imizamo Yethu. Those youngsters don't stand on the corners anymore ... There are so many people in Imizamo Yethu who appreciate what the patrollers are doing there ... they don't want the patrollers to stop.

He also explains that the SAPS, the CPF and patrollers have had, 'quite a number of meetings' and explained that patrollers cannot simply take the law into their own hands, they must also report crime to the station, which 'seems to be working now'.

While crime has reduced significantly and the patrols are now largely within the rule of law, there is little doubt that vigilantism did take place in Imizamo Yethu. The initial murders of the gang leaders, and several subsequent reports of excessive beatings (Hout Bay Organised Facebook site 2015/2016) point to informal (illegal) policing. Hence, the CPF Executive member (2016) commented, 'they went through a phase where they were beating people', and a member of the Hangberg CPF stated (CPF member 2016), 'they've got their patrollers beating them up over there'. There is certainly a murky line between non-violent neighbourhood watch and more informal, even illegal, means of controlling crime.

Informal policing in township and urban settings is widespread in South Africa and arises for a range of reasons. A long history of non-state policing in township settings is linked to the idea of people's justice (Buur & Jensen 2004). Indeed one respondent in Imizamo Yethu wistfully longed to return to 'the old ways' of 'sorting people out' when they committed a crime (Piper & Wheeler 2016). There is also a sense that the criminal justice system is not up to the task of effectively taking a criminal off the streets, and this prompts citizens to take the law into their own hands, often in violent ways (Bénit-Gbaffou 2008). In Hout Bay, many respondents identified the main problem as being that alleged criminals are given bail, known criminals get parole or early release or, if juveniles, are often returned to the care of their parents, and return to plague the community (SSP patroller 2016).

Informal governance may also thrive because, following the logic of 'anchored pluralism' police encourage 'community policing' that reduces their burden: 'Behind the official and sometimes ambiguous condemnation of vigilantism, the public enhancement of street patrols could be understood as a cynical encouragement of vigilantism, as a powerful form of community control of the social environment' (Bénit-Gbaffou 2008: 11). This attitude by the Hout Bay police was confirmed to us on several occasions (Community leader 2 2015). At times vigilante formations can be seen as 'twilight institutions', where it becomes difficult to distinguish unequivocally between what is state and what is not (Buur & Jensen 2004: 144). In Imizamo Yethu, the close relationship between the police, the CPF and the patrollers, which has developed since 2015, points in this direction. This shows both that the state does *not* have monopoly over the means of enforcing the rule of law, and that it implicitly endorses community attempts to reduce crime in ways that may violate constitutional rights.

Hangberg

Of all the areas reporting crime in Hout Bay, Hangberg has traditionally had the lowest statistics. While Hangberg may have lower levels of crime than Imizamo Yethu and the Valley, it is more likely that the community has a closed system of managing crime and little interest in involving the state or the SAPS. The governance of crime in Hangberg falls largely within the ambit of non-state actors, including both CPF/HBNW patrollers and those running organised poaching syndicates.

There are few murders reported in Hangberg, with the majority of crime linked to poaching, drug, and alcohol abuse (Hangberg youth 2016). Respondents noted that mostly they feel safe walking around at night in Hangberg and that

> most of the trouble is about poaching and drugs. ... If you snitched about me and my drugs at the cops then we will hurt you ... that kind of thing.

Figure 9.5 Youth in the street in Hangberg

> If you stand on my spot where I normally stand and sell my drugs … then there will be a fight.
>
> (Spaza 2016)

There are also incidents of domestic violence and child abuse. Poaching 'bosses' also pay some of their workers in drugs and 'use children to sell drugs on the street corners' (CPF member 2016). Children are also heavily involved in poaching syndicates (see Chapter 5) and there are regular cases where primary school children are tested positive for drugs.

There is a complex and troubling relationship between crime, safety, and security in Hangberg. It is not an easy scenario to resolve. Compounding long-standing problems with safety is the recent rise in visitors from Imizamo Yethu to Hangberg to avoid the new curfews enforced by the Imizamo Yethu patrollers. As a Hangberg CPF member (2016) explains, in Imizamo Yethu,

> during night time their shebeens are being closed … so they come this side to buy their alcohol here … right through the night. Even with stolen goods from their house-break-ins … they come here to sell it … because they can't walk in Imizamo Yethu at night. Because the patrollers beat them there if they find them.

This claim is verified by information from WatchCon tracking incidents of crime and suspicious activities (see Figure 9.6). There is a clear S-curve from Imizamo Yethu to Hangberg, indicating the housebreakings that may occur en route (SSP patroller 2016; CPF Exec 2016). In Hangberg, there is both little external governance assistance to combat social and security concerns, and a relatively low level of motivation to rectify them from within the community (Tefre 2010; SSP patroller 2016; CPF member 2016).

There is a long and contentious history between residents of Hangberg and state law enforcement. Chapter 4 describes the violence that took place in the 'Battle of Hangberg' that broke out between City law enforcement, the SAPS and the residents of Hangberg in 2010. The legacy of this urban battle has eroded the trust between residents of Hangberg, law enforcement agents (particularly from the City) and police. Numerous respondents describe how the SAPS cannot leave their vehicles unattended in Hangberg, as they will be vandalised (SSP director 2016; SSP patroller 2016; Captain 2016). There is also a widespread perception that certain officers in the SAPS are corrupt (Tefre 2010; CPF member 2016), which further undermines trust. While City-based law enforcement officials are still not welcome in Hangberg, the Hout Bay SAPS station claims to be slowly building trust in the community. The Hangberg CPF and the SAPS, for example, carried out a relatively successful joint neighbourhood watch operation in November 2016 (Spaza 2016; Captain 2016). It is clear, however, that the SAPS, has little legitimate authority, and a very limited role in providing safety and security in Hangberg. The state is struggling to 'anchor' the 'pluralism' of networked security governance; it is, instead, partly the 'dark nodes' that assert authority.

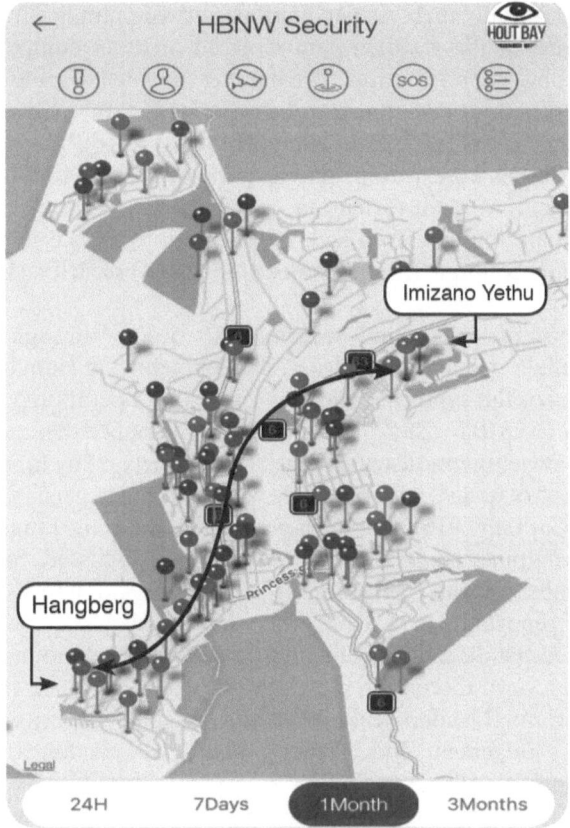

Figure 9.6 Trumpet data on incidents of crime for November 2016

Crime in Hangberg is policed by two informal means: through family ties, and through poaching and drug kingpins. Hangberg is a very close-knit community, with family links to the area going back generations. Three different young men living in Hangberg explained, 'We can look after ourselves here', 'We don't need a neighbourhood watch. ... We ARE the neighbourhood watch ourselves!' and 'Hangberg is too small ... everyone knows each other ... I can sleep outside when I'm drunk' (Hangberg youth 2016). The strong family ties and high levels of interconnectedness between families in Hangberg make it very difficult to report a crime even when the perpetrator is known (Tefre 2010). Thus, one crime prevention actor (SSP patroller 2016) noted that:

> It's a horrible thing to say but everyone's got a family member that's involved in something ... there's too many people that's got their fingers in the pie and then when there's an issue they blame SAPS, whereas they could say my aunty, my nephew, my niece is hiding drugs. I mean the police can only do so much.

Resolving crimes such as stealing or domestic abuse, in a context where police are not involved, usually involves negotiations between families. These discussions may include having a stolen item replaced or other compensation, or there may be physical retribution. The manner in which disputes are settled within the families does not appear to be organised in any particular way, and varies with each case (Tefre 2010; Hangberg youth 2016). This form of engagement reflects other observations on vigilantism, where 'More often than not, the emergence of vigilant formations is premised on a deep-seated mistrust of the police and/or perceived lack of initiative by police in providing basic human and economic (usually household goods) security' (Buur & Jensen 2004: 144).

Informal governance of safety and security also occurs through the authority of the four main poaching and drug-related kingpins in Hangberg. These individuals control the syndicates that run poaching operations, procure, and sell drugs in Hout Bay (SSP patroller 2016). The police themselves acknowledge that the poaching industry is large and supports many families in Hangberg: 'when you try to deal with poaching you are going to get a hiding from a lot of people' (Captain 2016). In a focus group with youth in Hangberg linked to one of the kingpins (Hangberg youth 2016), they discussed, rather chillingly, how a member of the Hangberg CPF who was kidnapped and nearly killed when he reported where drugs where hidden, would not be so lucky next time: 'He's gonna disappear again into the sea with lead around his neck'. Stating this in an open discussion illustrates the extent to which those with the backing of criminal leaders feel protected from state authority and their own community's judgement. Indeed they explain, 'we run things here ... [the cops] are scared of us all'.

In an anonymous recording by the relative of a kingpin (provided to researchers while conducting fieldwork in Hangberg), he explains the extent to which his relative controls crime, the wider Hangberg community, and the criminal justice system. He explains how this kingpin got his brother out of prison after serving only three weeks of a six month sentence; how he sells drugs to all sectors of Hout Bay (on such a large scale that he had to 'use a calculator'); how his relative murdered rival gangsters; where guns are stashed; the details of corrupt police, and more. The breadth of these activities highlights the power and authority of poaching and drug lords in Hangberg.

While there are individuals controlling poaching and drugs in Hangberg, there are no formal gangs in Hout Bay. There were established gangs in the 1990s in Hangberg but these were 'chased out' of the area by residents. Those involved in crime do have links to gangs from other areas of the Cape Flats, such as Hanover Park and Ocean View, specifically linked to the poaching industry, but gangs such as the Americans and the 28s do not operate in Hangberg (Captain 2016; Hangberg youth 2016). While the community is proud that there are no established gangs in Hangberg, the governance of safety and security in Hangberg is still deeply problematic. It is largely controlled informally, within kin networks, or by known criminals. There is very little governance from state agencies, CPFs, neighbourhood watches, or other forms of private governance. Indeed SSPs do not generally work or patrol in Hangberg (SSP director 2016).

The Valley

While informality is often associated with poorer communities, it is a mistake to assume it does not exist in wealthy areas too. In this regard, there is evidence that the governance of safety and security in the Valley does include informal activities that contradict City by-laws. For example, the rise of private security governance produces informality in the form of illegal road closures and booms. 'Tarragona' and 'Longkloof', two sectors deep in the Valley, have put up a road boom with a guard at their entrance: 'it's not really a gated community, but there is a security guard at the boom ... if somebody looks suspicious the security will ask a few questions' (CPF Exec 2016). Mount Rhodes, the area where the murder of a tourist that triggered the formation of HBNW occurred, is also privately enclosed. Members of the area gathered donations from about 70 per cent of the residents, then erected palisade and electric fencing around their entire sector, and put a guard and CCTV monitoring at the entrance (Tefre 2010). These areas are public spaces and the fences and booms are on council land, and thus illegal. As one SSP explained (SSP director 2016), 'we put up a boom. Lots of laws say you are not allowed to but we never took it down'. In fact, ADT refused to erect the boom, as they did not want to be seen guarding an illegal boom and feared a lawsuit.

The council and law enforcement agencies have effectively ignored this informal governance of crime. In terms of Mount Rhodes,

> [They got] kind of a guarded consent from the municipality, you're not really allowed to do it, but they had got a murder to point at and say 'well, what are you going to do about it?' So the municipality has let it slide because obviously they can't give permission as it goes against the constitution.
>
> (A previous Chair of HBNW in Tefre 2010: 99)

As Roy (2005: 233) might put it, these 'unauthorised' constructions can be seen as 'expressions of class power ... that come to be designated as "formal" by the state while other forms of informality remain criminalized'. Indeed, while illegal, these actions would unlikely be labelled a 'dark node' of security governance.

Booms and suburb closures are not the only ways in which the Valley enacts informal governance. An SSP provider also explained how he assisted Valley residents who live in an area adjacent to Imizamo Yethu to lock a side gate that provided access to the Valley. There was a popular backlash against this informal attempt to control access, as residents of Imizamo Yethu broke the lock and made sure that the gate stays open (SSP director 2016). As well as indicative of ineffective governance by the local state, private and informal control over public space highlights the extent to which the neo-liberal system of private property rights colonises public spaces, and the extent to which private capital, rather than the City, influences the day-to-day governance of crime and security.

As discussed above, HBNW supports a network of 20 to 30 'first responders'. Their activities vary widely and largely fall within the bounds of the law. While it is legal to attend a crime scene without the police, there is a 'grey zone' of what citizens are allowed to do in regard to safety and security. First responders have, at times, been characterised as 'gung ho', as there have been instances of armed civilians chasing suspected criminals. The police argue that civilians should be the 'eyes and ears of the police and not get directly involved, but as is clearly demonstrated by parts of the HBNW, not all are willing to settle for this passive role and take on a more direct approach' (Tefre 2010: 112). Indeed, one previously active member of the CPF, working in a neighbourhood watch capacity, is well known for apprehending and detaining suspects in all areas of Hout Bay. He described how he approaches a suspect:

> I will treat him according to the constitution and make a citizen's arrest. I will restrain him, put him on the floor and cuff him, but it needs to be a schedule one offense ... we are a lot more hands on as opposed to eyes and ears.
>
> (SSP patroller 2016)

He also described more legally contentious acts when explaining,

> So last night I find [names two known criminals] and I ask them 'guys where are you going?' and they say 'we are going to the harbour'. 'What do you have on you?' They had drugs and knives on them, leaving IY on their way to the robots. So I take the stuff from them. They turn around and go back home. Now I can't arrest them, it's an apple/biltong knife. I'll just waste the court's time.

Reframed, this could be a description of theft of personal possessions as there was no evidence of a crime being committed. However, this form of preventative action is strongly supported by most members of the community, including the police. The same respondent described how at times he will deliver a suspect to SAPS cells, 'if it's a direct instruction by SAPS' even though legally SAPS must transport all suspects from the site of a crime (SSP patroller 2016; SSP director 2016).

The informal governance of safety and security in the Valley arises for many of the same reasons we see acts of vigilantism in Imizamo Yethu: '[I]f the police are not really in a position to defend us, then we need to defend ourselves. And it's a basic right, the civil right of self-defence' (Valley respondent in Tefre 2010: 101). Indeed the SSP patroller (2016) explained how frustrated he is with the criminal justice system and how this legitimises working, at times, outside of the Constitutional framework:

> See these two guys [shows us a picture of suspects]. They've been robbing houses since they've been 12 years old. They are both 28s in prison, terribly violent, they've stabbed people, beaten people, raped people. ... So half past four in the morning I catch them in the flowerbed of an old couple, balaclavas on, gloves on, knives, everything. They go to court in the morning. Two days later and the court ask them 'why did you have balaclavas and gloves on?' They said 'we were cold'. 'Why did you have knives?' 'We needed to protect ourselves'.

The respondent described how these suspects were released with no charges, how this is a frequent occurrence, and how their history cannot be used in court, and thus concluded, 'the problem is the Constitution'.

The consequences of contending forms of safety and security governance

This research shows, as does much of the literature on community policing and private security (Diphoorn & Kyed 2016; Bénit-Gbaffou 2008), that it is not the state or the SAPS that influences the day-to-day governance of security in areas such as Hout Bay. Rather, non-state, nodal actors including businesses, civil society formations and residents themselves co-construct various and sometimes contending forms of security governance in Hout Bay. Analysis of the different forms of governance in urban settings, in this case bureaucratic, developmental, market, network, and informal, highlights the limited power of the state over day-to-day urbanisms. Indeed, this survey of safety and security governance shows how the power, legitimacy, and thus authority of non-state nodal actors is expanding.

However, this is not to equate this pattern to an automatic decline of state power. The state still, to some extent, anchors the pluralism of network and nodal security governance in Hout Bay. For instance, the privatisation of security does not necessarily lead to an erosion of the state, nor is it automatically a symptom of state failure (Diphoorn & Kyed 2016). It may be a reflection of the neo-liberal idea that to 'govern less is to govern better', or alternative notions of network or nodal governance where the state may be 'steering' safety and security governance rather than 'rowing' (Abrahamsen & Williams 2007: 132). In the context of Hout Bay, the state is still central to security governance in relation to arrests and the related criminal justice procedures. However, in terms of crime prevention and security provision more generally, the state is just one oarsman in a boat with many rowers. The state thus has control over the prosecution of criminals, but the prevention of crime is a collective effort.

The magnitude of the authority of the neighbourhood watches, private security companies, patrollers, and crime kingpins is a clear indication of limited state power in regards to crime prevention. Ilustrations include: how the HBNW paid for a City-branded vehicle for volunteer law enforcement officers (a move approved of by the City); how the CPF have financed a cell-phone for an SAPS Hout Bay station Captain; how CCP are usually on the scene before the SAPS in regards to crime in the Valley; how respondents in Imizamo Yethu felt that those who make the community safe are 'cats and dogs' because cats 'kill rats and mice that eat food', and dogs 'bark at tsotsis'. These instances highlight the limited sovereignty and authority of the police (HBNW Exec 2016; CPF Exec 2016; Piper & Wheeler 2016). Indeed, an HBNW Executive member (2016) stated that, 'The neighbourhood watch can't fix the police so we need to create a private police force. We need to provide manpower to assist the police.'

Even more remarkable are the stories of how private actors attempt to control police activity. One community 'crime-fighter' explained that he calls a captain at the Hout Bay station 'more than I call my wife', and is able to generate quick responses from the SAPS (SSP patroller 2016). In a different context the SAPS is, at times, not even able to enter Hangberg. This mirrors Steinberg's work on police ineffectiveness where he argues that, 'it is the citizenry who determine to what extent they are policed'. The police perform in a context where the audience (the community) dictates the script and if the police do not perform correctly, the community will become the lead actors (Steinberg cited in Tefre 2010: 21). It must be noted, however, that the SAPS, in their attempt to symbolically anchor their authority, are not averse to forming partnerships with private actors, even while they would like to maintain more control than they do. The logic is entirely pragmatic. They simply cannot do it on their own.

Spatial and racial consequences of intersecting forms of urban governance

There are undoubted consequences to the part privatisation, socialisation, and informalisation of safety and security in Hout Bay. First, both civil society-led governance and market governance may lead to the depoliticising of issues of crime and safety. Debates on community safety may too easily become narrowly framed, or in the case of Hout Bay, spatially- and racially-based. We witnessed the narrow framing of safety and security issues at a CPF meeting held one evening to provide feedback on crime prevention after the spate of attacks linked to gangsterism in Imizamo Yethu. The event was well attended, with hundreds of residents, mostly from the Valley, present in a packed school hall. However, another CPF meeting was taking place at the same time in Imizamo Yethu. The separation of these meetings implies there is no clear sense that security issues are connected, and that Hout Bay is one community. The Valley needed to discuss 'their' problems, while Imizamo Yethu addressed its 'own' issues, despite the fact that they were discussing the

very same problem. Several members of the audience questioned this separation but the majority view was that there was no need to integrate the
discussions on safety and security.

Certainly, market governance, particularly linked to private companies,
tends to be depoliticised altogether. On this logic, residents are rendered individual consumers who choose where and how to manage crime, in significant part, based on what they can afford. This resonates with the work of
European scholars of network governance who have

> presented network governance as a fundamental facet of neoliberal
> hegemony ... Rather than the development of new plural, horizontal and
> inclusive forms of network governance, critics say, what we observe in
> European cities is the increasing concentration of urban power in the
> hands of a few political and business elites.
>
> (Blanco 2015: 124)

Security becomes an individual commodity rather than a public good. Linked
to this concern is the fact that private actors do not have to take broader community concerns, such as unemployment, into account. Both Deep Blue and
HBNW described how they are not interested in having patrolling security
guards on their books. Deep Blue explained that it is becoming too expensive
as 'the government want to make security guards part of the middle class and
we can't afford that' (SSP director 2016). For HBNW, hiring security guards
would come with too many regulatory hassles (HBNW Exec 2016). For both
organisations, cameras are an obvious replacement for human personnel;
cameras do not demand the minimum wage and they can be more reliable
and placed in more areas.

Lastly, but significantly, private and informal governance of safety and
security also leads to reinforcing patterns of neo-apartheid segregation. South
Africa remains a strongly divided society, where citizens believe their social identities, cultural values, and material interests are in conflict with those of others
around them (Bradford, Huq, Jackson, & Roberts 2014). The actual practice
of policing Hout Bay reflects closely Said's notion of 'other' as an invented
concept, where a social group constructs the identity of the 'other' social group
only in relation to themselves, rather than as a separate entity (Lemanski 2004).
As Pillay (2008: 149) explains, 'the existence of the apartheid state, and the
mobilisation against it, created multiple socially cohesive communities around
the binary of domination and resistance'. Certainly, the largely white community of the Valley feels a fear of domination by criminals that at times seems
reminiscent of the apartheid concept of 'swart gevaar' (black danger). There is
a shared identity of being victims of crime, where having 'Imizamo Yethu in
their midst seems to have reinforced a feeling of "us" in the Valley falling prey
to "them", the criminal elements of Imizamo Yethu' (Tefre 2010: 101).

While no respondents in this research made overtly racist comments, there is
clear racial profiling evident in security practices in Hout Bay. Thus, patrollers

in Imizamo Yethu made a joke on the HBNW radio that there was a 'suspect person' in the township, when a white member of the CPF was walking around with them. This is a clear reference to the racial profiling that occurs in the Valley. The manner in which patrolling takes place is further reinforcing neo-apartheid spatial patterns. While technically there is one CPF and one HBNW for all three areas, in practice different people patrol the areas, and there is little mixing between the areas. This reinforces the perceptions of 'other' and does little to break down racial barriers. Those that do work across all three areas, such as the leader of CCP, are more aware of the root causes of crime (SSP patroller 2016). There is, however, one positive consequence of private governance breaking down apartheid social structures: the closer relationship between some members of Imizamo Yethu and Hangberg since Imizamo Yethu residents have been forced to pursue drinking and associated illegal activities in Hangberg.

Conclusion

This chapter has demonstrated that the governance of security and safety in Hout Bay is shared between civil society, private companies, informal actors, and the state. Further, in stark contrast to the governance of property outlined in Chapter 3, and of livelihoods in Chapter 5, developmental, network, market, and informal governance operate side-by-side with relative coherence in the case of crime prevention in Hout Bay. (The state alone deals with the prosecution of suspects and punishment of criminals – aside from rare incidences of mob justice.) However, these multiple forms of security governance do not address the causes of crime, nor the neo-apartheid forms of segregation that characterise the settlement. These facts affirm the larger point of the book: that the co-existence of a framework of multiple and contesting forms of urban governance (FUG), whether in respect of property, livelihoods, security or any key urban need, reflects the limited power of City Hall to construct the urban landscape on its own. Indeed, this incapacity is most pronounced in respect of what, since Hobbes, many hold as the core function of the state: the provision of safety and security. Consequently, local democracy, formally linked to City Hall, is disconnected from many forms of governance that produce the urban. In addition, network, developmental, market, and informal forms of governance work in terms of logics of connection, money, or kin rather than of civil equality. Thus, both structurally and systemically, democracy is disconnected from urban governance in Hout Bay.

Notes

1 Chairpersons of CPFs in an area sit on area boards. Representatives of these area boards then sit on provincial boards. At a national level, the nine representatives of the provincial boards sit on a national Community Policing Consultative Forum (Mottiar & White 2003).

2 There is a third company, SMA, working as an SSP in Hout Bay, but it is much smaller than the other two and focuses on providing security at events or in specific locations, rather than on armed response (SSP director 2016; Captain 2016).

References

Abrahamsen, R., & Williams, M. (2007). Introduction: The Privatisation and Globalisation of Security in Africa. *International Relations*, *21*(2), 131–141.

Bénit-Gbaffou, C. (2008). Community Policing and Disputed Norms for Local Social Control in Post-Apartheid Johannesburg. *Journal of Southern African Studies*, *34*(1), 93–109.

Blanco, I. (2015). Between Democratic Network Governance and Neoliberalism: A Regime-Theoretical Analysis of Collaboration in Barcelona. *Cities*, *44*, 123–130.

Boutellier, J. C., & van Steden, R. (2011). Governing Nodal Governance: The 'Anchoring' of Local Security Networks. In Crawford, A. (Ed.), *International and Comparative Criminal Justice and Urban Governance: Convergence and Divergence in Global, National and Local Settings*. Cambridge: Cambridge University Press, 461–482.

Bradford, B., Huq, A., Jackson, J., & Roberts, B. (2014). What Price Fairness when Security is at Stake? Police Legitimacy in South Africa. *Regulation and Governance*, *8*(2), 246–268.

BusinessTech. (2016). Cape Town is now among the 10 most violent cities in the world. *BusinessTech*. https://businesstech.co.za/news/general/110133/cape-

Buur, L., & Jensen, S. (2004). Introduction: Vigilantism and the Policing of Everyday Life in South Africa. *African Studies*, *63*(2), 139–152.

Diphoorn, T., & Kyed, H. M. (2016). Entanglements of Private Security and Community Policing in South Africa and Swaziland. *African Affairs*, *115*(461), 710–732.

Genever, S. (2015). Mob wants justice after foreign national's death. *South Africa Breaking News*. www.sabreakingnews.co.za/2015/07/10/mob-wants-justice-after-uct-students-death/. Accessed on 12 October 2016.

HBNW website (2016). www.houtbaywatch.com/index1.php?Link=45&Article=45 HBNW welcome letter.

Hosken, G. (2016). Fear rules Joburg, but Cape Town is murder capital. *Times Live*, 29 June. www.timeslive.co.za/sunday-times/news/2016-06-29-fear-rules-joburg-but-cape-town-is-murder-capital/ Accessed 10 December 2016.

Hout Bay Organised Facebook site (2015/2016). www.facebook.com/groups/houtbay/.

Irish, J. (1999). Policing For Profit: The future of South Africa's Private Security Industry. www.africaportal.org/publications/policing-for-profit-the-future-of-south-africas-private-security-industry/. Accessed 8 May 2018.

Johnston, L., & Shearing, C. (2003). *Governing Security: Explorations in Policing and Justice*. London: Routledge.

Kriegler, A., & Shaw, M. (2016). Facts show South Africa has not become more violent since democracy. *The Conversation*. http://theconversation.com/facts-show-south-africa-has-not-become-more-violent-since-democracy-62444. Accessed on 12 December 2017.

Legg, K., & Kalipa, S. (2015). Boy, 16, dies in spate of attacks. *IOL Online*, 27 July. www.iol.co.za/news/crime-courts/boy-16-dies-in-spate-of-attacks-1891255. Accessed 8 May 2018.

Lemanski, C. (2004). A New Apartheid? The Spatial Implications of Fear of Crime in Cape Town, South Africa. *Environment and Urbanization, 16*(2), 101–111.

Loader, I., & Walker, N. (2006). State of Denial? Rethinking the Governance of Security. *Punishment and Society, 6*(2), 221–228.

Mottiar, S., & White, F. (2003). *Co-Production as a Form of Service Delivery: Community Policing in Alexandra Township.* Johannesburg: Centre for Policy Studies.

Pelser E. (1999). The challenges of community policing in South Africa, Occasional paper no. 42, *Institute for Security Studies.* Pretoria: ISS.

Pelser, E. Schnetler, J., & Louw, A. (2002). *Not Everybody's Business: Community Policing in the SAPS' Priority Areas.* Pretoria: ISS.

Pillay, S. (2008). Crime, Community and the Governance of Violence in Post-Apartheid South Africa. *Politikon, 35*(2), 141–158.

Piper, L., & Wheeler, J. (2016). Pervasive, but not Politicised: Everyday Violence, Local Rule and Party Popularity in a Cape Town Township. *SA Crime Quarterly, 55,* 31–40.

Plato, D. (2013). Police must Explain Disastrous Police-to-Population Ratios: Statement by Dan Plato, Western Cape Minister of Community Safety, 10 July. www.westerncape.gov.za/news/police-must-explain-disastrous-police-population-ratios. Accessed on 10 October 2016.

Raab, J., & Milward, H. B. (2003). Dark Networks as Problems. *Journal of Public Administration Research and Theory, 13*(4), 413–439.

Raba, B. (2016). Cape Town's police-to-population ratio way below national norm. *The Sowetan.* www.sowetanlive.co.za/news/2016-07-27-cape-towns-police-to-population-ratio-way-below-national-norm/.

Rosenbaum, D. P., & Lurigio, A. J. (1994). An Inside Look at Community Policing Reform: Definitions, Organizational Changes, and Evaluation Findings. *Crime & Delinquency, 40*(3), 299–314.

Roy, A. (2005). Urban Informality: Toward an Epistemology of Planning, *Journal of the American Planning Association, 71*(2), 147–158.

SANParks. (n.d.). Table Mountain National Park. SANParks. www.sanparks.org/parks/table_mountain/conservation/marine.php. Accessed 15 December 2017.

SAPS. (n.d.). Crime statistics: Integrity. SAPS. www.saps.gov.za/services/crimestats.php. Accessed 8 December 2016.

town-is-now-among-the-10-most-violent-cities-in-the-world/. Accessed 10 December 2016.

Tefre, Ø. S. (2010). Persistent inequalities in providing security for people in South Africa-A comparative study of the capacity of three communities in Hout Bay to influence policing. Master's thesis, University of Bergen.

Interviews

1. Community leader 2. (2015). Interviewed by Fiona Anciano and Laurence Piper. 17 August 2015.

2. CPF Exec (CPF executive member). (2016). Interviewed by Fiona Anciano. 23 November 2016.

3. CPF member (2016). Interviewed by Fiona Anciano and Conrad Meyer. 23 November 2016.

4. Spaza (Hangberg Spaza shop owner). (2016). Interviewed by Conrad Meyer. 26 November 2016.

5. Hangberg youth. (2016). Focus group with youth in Hangberg. Conducted by Conrad Meyer. 27 November 2016.

6. HBNW Exec (HBNW Executive member). (2016). Interviewed by Fiona Anciano. 7 December 2016.

7. Captain (Hout Bay SAPS). (2016). Interviewed by Fiona Anciano. 7 December 2016.

8. Pastor 1. (2015). Interviewed by Fiona Anciano and Conrad Meyer. 24 November 2016.

9. Patrollers. (2016). Focus group with Patrollers in Imizamo Yethu. Conducted by Fiona Anciano and Conrad Meyer. 23 November 2016.

10. Poachers. (2016). Focus Group by Fiona Anciano and Conrad Meyer. 17 May 2016.

11. SSP director. (2016). Interviewed by Fiona Anciano. 6 December 2016.

12. SSP patroller. (2016). Interviewed by Fiona Anciano. 23 November 2016.

Conclusion

Democracy, governance, and neo-apartheid

Democracy is more in demand and supply than ever in human history, and yet dissatisfaction with democratic rule endures (Norris 2011). Hout Bay, our case study site, captures this sentiment of democratic deficit almost exactly. Despite active engagement through democratic means, including new participatory spaces, most residents feel frustrated with the response of the state. The key reason for this is the weak connection, and even disconnection, of local democratic institutions from actually existing forms of governance. City Hall is often a junior partner in, or absent altogether from, the many forms of governance that shape Hout Bay. Consequently, the institutions of local democracy have little or no purchase over key forms of power that control and constitute the City of Cape Town.

This weak connection between democratic institutions and urban governance is not the only problem that confronts democratic rule in the urban South, but it is the one most often ignored. Typical responses to the problem of democratic deficit focus on reforming democratic institutions or on enhancing the conditions for democratic politics. As confirmed in our case, while these are important issues, they are less significant than the impotence of elected city officials (otherwise termed representatives or politicians) in relation to the multiple and contending forms of governance that rule the city. Indeed, under conditions of rapid social change, such as urbanisation by poor migrants, the primary challenge that confronts rule in the Global South is amassing the productive power necessary to create the city. In the gap between urban poverty and productive power, informality flourishes.

We began our case by characterising the mainstream model of democratic rule as a virtuous circle of responsiveness between politics, democracy, and government (see Figure C.1). Politics involves individual residents and groups forming views, and organising and acting on issues that they feel need addressing. Democracy refers to the institutions (civil and political rights, elections, and organs of state) that facilitate the election of key officials into government, and that help residents to influence officials' decision-making between elections. Last, but not least, government refers to the making and implementing of important decisions by elected officials, rather than the

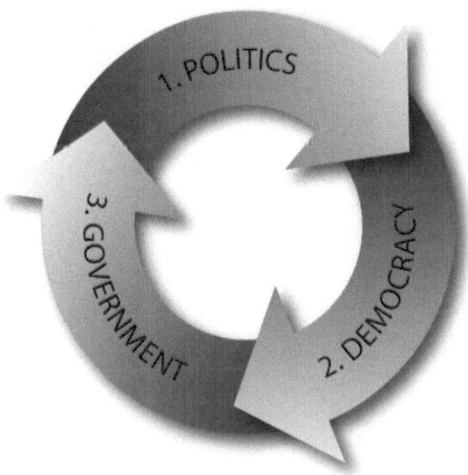

Figure C.1 Model of democratic rule

security establishment or the bureaucracy. This closes the feedback loop in democratic rule, linking government to politics via democratic means.

Perhaps our key insight is that, when viewed from the perspective of residents of the city, decision-making over urban life is not monopolised by City Hall, especially the elected officials of City Hall, but rather it is shared among a multitude of actors including national and provincial government, business, and residents themselves. Further, by focusing on the actual experience of rule, we surface many forms of governing distinguishable not so much in terms of 'who governs' but 'how' they govern; that is, sets of rules that embody distinct logics or rationalities. These forms of governance are contextually defined, but, in our case, include the bureaucratic government of City Hall; the developmental governance of the national and provincial state and philanthropic donors; market governance where business meets individual wants through economic exchange; and nodal governance where a range of actors from state, business, and society co-operate to solve problems, such as safety and security. Other kinds of governance also exist, including informal types of governance that influence how the poor live outside the rules of the formal system.

In addition to identifying a multiplicity of forms of governance (see Figure C.2), we point to their often-contradictory logics, and how these intersect in 'topological' fashion to produce a contingent and emergent framework of urban governance (FUG). This yields three further observations. First, most forms of governance imagine political subjectivities that potentially fragment democratic citizenship. Second, different places in the city are characterised by different combinations of governance, segregating the city in political as well as social terms. Third, the topological nature of governance differentiates the city

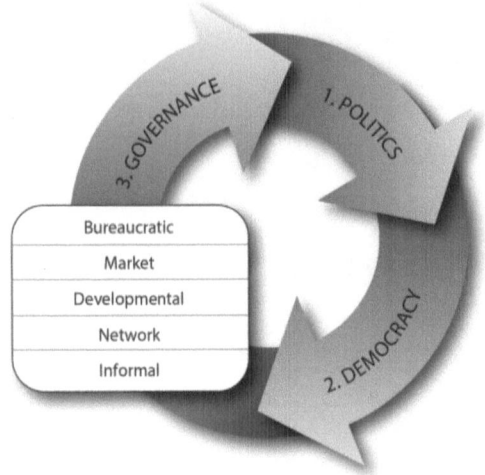

Figure C.2 Actual local democratic rule

into well-ordered, and unevenly ordered places. In South Africa, this undoing of the demos takes an enduring racialised form that we term neo-apartheid.

In what follows, we reflect on the implications of the particular framework of urban governance for democratic rule in Hout Bay, making the case that while there are some issues in respect of politics, and more in respect of democracy, the biggest challenge concerns the limited power of elected officials over important forms of decision-making that shape the City.

Assessing democratic rule in Hout Bay

Our assessment of democratic rule in Hout Bay follows in sequence the three elements in the virtuous circle of politics, democracy, and government. To aid our analysis of the potential exclusion of the urban poor from politics and democracy in Hout Bay we draw on Steven Lukes' (2005) conception of power as control over people, or as domination. We argue, using Lukes' three-fold model, that democratic politics is undermined, and domination is on the rise, when forms of a) repression, b) manipulation, or c) depoliticisation emerge. In respect of government and democratic rule, we draw on Stone's (1989) conception of power as social production to assess the capacity of elected officials to constitute the City.

Politics and belonging

Politics is defined conventionally as the expression of public views on key issues that affect residents, the formation of organisations to address these views, and the prosecution of actions by these organisations and residents to bring about change. In a democratic society, these capacities are expressed as civil and political rights, the dispossession of which is taken to be some form of domination. As the wider literature on the Global South suggests, the group most vulnerable to domination is the urban poor, or subaltern groups.

In general, we found little by way of repression, manipulation, and depoliticisation of politics in Hout Bay. In respect of repression, there have been many moments of coercion by the state to stop illegal or violent protests by poor, black residents. Well-known examples include the Battle of Hangberg in 2010, the 2011 protests against taxi formalisation, the protest against 'super-blocking' in Dontse Yakhe in 2017, and the violent protests against fishing quotas most recently. However, in all of these cases, the actions of the police can be defended as attempts to restore order when protest was no longer peaceful. In this respect, the coercive action of the state was consistent with the principles of democratic politics. Indeed, it is clear that all residents of Hout Bay see protest as a common phenomenon that is part of democratic rule, and all major political parties and civil society formations have engaged in protest at some point. Notably, poor, black residents lead most protests in Hout Bay, confirming that there is room for vulnerable groups to engage in disruptive mobilisation.

An important form of repression, common under apartheid, was the lack of a diverse and independent mass media free from state censorship. This clearly no longer applies in South Africa, which has a diverse, healthy, and free media, albeit issues of the disproportionate bias of 'the view from the suburbs' persist. Indeed, as we argue elsewhere (Piper et al. 2017), this same tendency is evident in the *Sentinel*, the free weekly community newspaper for Hout Bay. Importantly, the *Sentinel* is not the only newspaper in Hout Bay, and anyone is free to issue rival publications, a good example being *Hout Bay Speak* put out by ANC leaders (Piper, von Lieres, & Anciano 2017). Perhaps even more important is the rise of social media, for example the huge Hout Bay-specific user groups on Facebook, where the views of Hangberg and Imizamo Yethu residents are a little more evident. This granted, challenges with media access and bias endure.

In terms of organisation and action, politics in Hout Bay is largely inclusive. A significant diversity of political parties exists, with the ANC and DA rooted in the major settlements of Imizamo Yethu and the Valley respectively. In addition, political parties like the Economic Freedom Fighters (EFF) and the African Christian Democratic Party (ACDP) also have a presence, and the ANC and DA are able to campaign in the strongholds of their opponents with some qualifications. Furthermore, although SANCO attempts to monopolise representation of Imizamo Yethu to the state around development processes, this claim is contested both by other factions of leadership and by rival organisations and groups that emerge periodically. The fact that the ANC office and homes of key ANC leaders were destroyed in 2017 reflects the potential (over) assertiveness of residents.

The contestation of community leadership is even more evident in Hangberg. Indeed, so intense was this that, after the Battle of Hangberg, the state instituted the PMF to quell local rivalries. While good for peace, the PMF has had parlous consequences for democratic representation. Nevertheless, there remain a substantial number of civil society formations based, or active, in Hangberg despite the PMF. Lastly, while the Hout Bay Ratepayers and Residents Association (HBRRA) tends to dominate the view from the Valley, numerous other civil society formations are comfortable articulating independent views. In addition, there any many NGOs that work daily across Hout Bay from the environmental organisation THRIVE, to philanthropic church formations such as St James House and animal shelters such as the Domestic Animal Rescue Group. In short, in both organisational and action terms, civil life in Hout Bay is extensive, diverse, active, and often contentious.

Perhaps the major constraint on democratic politics in Hout Bay comes in the form of popular discourses of belonging that inhere in political talk. Thus, while almost everyone we spoke to supports democracy, inherent in the several approaches to politics are divergent conceptions of belonging linked to identity that undermine the idea of equal citizenship, and therefore the right of all residents to participate in political life. The effect of invoking identity and belonging to delegitimise residents' participation in local politics is a form of depoliticisation. It speaks directly to Lukes' (2005) arguments about forms of power that shape the very preferences of people to conceal power relations that harm their interests.

The two most frequent discourses we encountered were those of 'racial transformation' and 'conservatism as conservation'. Racial transformation is the discourse common among the ANC, SANCO, and leaders in the black community more widely, that links democracy to uplifting the historically oppressed majority through priority attention from the developmental state. This discourse gives a racial and partisan twist to the biopolitical governance of populations that Chatterjee (2004) argues is common in the Global South. Thus, it is the responsibility of the state to advance the marginalised, who in this framing are the historically oppressed black majority. Liberation nationalism tends to conjoin the black majority with both the South African nation and the ANC political party (Piper 2015; Piper & Anciano 2015). This is most evident in Hout Bay around the politics of housing in Hangberg and Imizamo Yethu.

Racial transformation is thus interpreted in a nationalist sense, manifested in the attempted pogroms against foreign residents of Imizamo Yethu in 2008 and 2009. Racial exclusion is also central to the conflict between SANCO and the HBRRA over the development of Imizamo Yethu described in Chapter 6, with partisan politics mapping closely onto this racial divide, although to a lesser extent. Consequently, divides over the PMF leadership in Hangberg are sometimes interpreted through a party-political lens, although all concerned are black people. The point is that these discourses articulate a hierarchy of belonging that makes it more difficult for foreign nationals, white people or the DA to engage in politics in Imizamo Yethu on the one hand, or black people and the ANC to do likewise in the Valley, on the other.

'Conservatism as conservation' is the approach of the Hout Bay Ratepayers and Residents Association (HBRRA), and an associated network of older, white professionals, that are opposed to both the market-friendly densification agenda of the City, and the influx of poor, black migrants into Hout Bay. Invoking notions of the 'historical character' of Hout Bay and 'conservation of the environment', this group has mobilised to prevent the expansion of Imizamo Yethu, the densification of Hout Bay through greater property subdivisions (Chapter 6), and the privatisation of Chapman's Peak Drive (Chapter 8). Notably, conserving the natural environment from human encroachment has become the unifying theme behind this approach.

Importantly, while both of these sets of discourses make political engagement in Hout Bay more fraught, they do not make it impossible. For example, the nationalist exclusion in the 'racial transformation' discourse is challenged by refugee rights NGOs that articulate a human rights approach to belonging. In addition, some groups in Hangberg have challenged their racial marginalisation of 'not being white enough under apartheid, and not being black enough today' through articulating a KhoiSan nationalism that trumps race with a claim to an autochthonous San ethnicity. Others have chosen the route of Rastafarianism that affirms a similarly prior African identity. Furthermore, the conservationism of the HBRRA is offset by a more radical ecologism, evident in NGOs like THRIVE who, rather than excluding humans from nature, affirm the integration of human and non-human systems in a sustainable way that also addresses social justice. On balance, then, there are diverse and contending political discourses of belonging in Hout Bay, indicative of an open political environment.

Democracy and responsiveness

In assessing the health of democracy in Hout Bay, we focus on both representative institutions such as elections, the ward councillor and subcouncil 16, and participatory institutions such as the ward committee, and public consultations on the budget and on various development projects. The literature on democracy in the urban South suggests that the closure or capture of democratic spaces by local elites is the main threat to institutions. Once again using Lukes' theory of power, we explore the democratic institutions of Hout Bay in terms of repression, manipulation, and depoliticisation. In short, the evidence shows that democratic institutions are widely used in Hout Bay, and that representative institutions work much as intended. However, some issues of manipulation arise that reflect a constricting of democratic space, and that partly explains the unresponsiveness of elected officials.

In terms of repression, we found little evidence of elites overtly shutting down representative or participatory structures. Hout Bay has had 10 local and provincial/national elections since 1994, all of which were free and fair.

Contests over elected office in Hout Bay are closely fought, with the DA coming out on top of the ANC only in recent times, mostly due to the switch of Hangberg voters from around 2007. Furthermore, the key constituencies all turn out in good numbers at election time, albeit the DA is about 5 per cent better at getting its voters to the polls than the ANC. Electioneering, too, is largely free and fair, although the campaigning of the DA in Imizamo Yethu proceeds under some sufferance. In short, there is little evidence for the exclusion of the urban poor from electoral participation.

Similarly, there is little evidence of the repression of participatory spaces. Thus, whether regarding the activities of the ward committee, or consultations on housing projects, fishing policy, the tolling of Chapman's Peak Drive, or most recently the location of the new clinic, most affected groups have engaged in these spaces when the opportunity has been presented. More significant in this respect are attempts by powerful civil society groups to capture or dominate democratic spaces. Key examples here include the allegations that the DA captured the PMF in Hangberg; that SANCO excluded rival groups from participating in the housing PHP process in Imizamo Yethu; and that Entilini co-opted key leaders from Imizamo Yethu and Hangberg onto the Chapman's Peak Development Trust to demobilise opposition to the Toll Road.

While at times these attempts have proved successful, they are always vulnerable to contest by a rival faction (as with SANCO and the HBRRA), or an entity from a rival social group (as with super-blocking). The only exception to this narrative is the decision to monopolise community representation on development projects in Hangberg through the Peace and Mediation Forum, a decision that has created an opportunity for new forms of patronage while also constricting the space for civil society.

Perhaps the most pervasive form of the manipulation of democracy is state officials' cherry-picking of the views expressed in consultative forums. Cherry-picking is manipulative in that officials can choose views or positions in contested forums that suit them, rather than those that confront them. While it can be difficult for officials to respond satisfactorily when public views are deeply divided, there is also strong evidence of officials ignoring majority views in the legally required consultation on the Toll Road, for example. Similarly, a recurring theme in the debate over housing in Imizamo Yethu is the choice of City officials to ignore the majority view of residents for more housing in favour of developing settlement-specific facilities. Perhaps most profound of all is the frustration of fisher folk in Hangberg who have been consistently ignored or placated in their quite reasonable demands for licences.

A form of manipulation similar to cherry-picking is the unresponsiveness of elected officials to residents' engagements, whether through representative or participatory processes. Examples include the lack of support by a former ward councillor in Hout Bay for the *in/formal south* project, or in backing popular demands for housing in Imizamo Yethu, or for fishing licences in Hangberg. A common view of respondents is that the former ward councillor was responsive only to her constituency, the residents of the Valley, and ignored others. This is a version of what Americans term 'pork barrelling', where politicians secure 'spending on projects and programmes to benefit specific party constituents' (in Ledeneva 2018: 349). As McClendon (2016) points out, this behaviour is typical of politicians in South Africa, and we suggest reflects the electoral logic where politicians first service the preferences of their supporters. In addition, elected officials on subcouncil 16 were reluctant to support civil society initiatives from below in Hout Bay, either due to a lack of power to get appointed officials to work across silos, or due to a fear of stoking unnecessary conflict in Hout Bay.

Notably, the avoidance of disruption was a recurring theme in government approaches to rule in Hout Bay, with the outcome that doing nothing was often a preferred choice for officials than doing something. This response is born of hard lessons. But, when engagement and participation fail, community organisations and members turn to disruption through mobilisation to force accommodation. Examples include the protests against fishing licences, super-blocking, and the Toll Road. Conversely, elites such as City officials or project managers, turn towards the co-option of community leaders onto representative forums to prevent conflict. Examples include the incorporating of taxi owners into the BRT companies to prevent violence, and co-opting community members into the Chapman's Peak Development Trust to forestall disruption; but the exemplar is the Peace and Mediation Forum established after the Battle of Hangberg.

Lastly, we found some evidence of depoliticisation in the neo-liberal managerialism from the City of Cape Town, expressed in its vision to become a 'world class city'. Neo-liberal managerialism reads democracy through the lens of business-friendly policies for economic growth combined with the 'biopolitical' management of the poor. The latter discourse interprets the problems of Hout Bay overtly in class rather than racial terms, as in the talk of no place for crowded, informal, and poor settlements on the slopes of the mountain. This framing of democracy resonates with Wacquant's (2010) notion of a Centaur state that entails freedom for the elite and middle classes, but punitive regulation of the poor. In South Africa, there are also racial undertones to the management of poor populations, where some associate poverty and other social ills with blackness. Importantly, though, this politics is offset by the DA's wariness of both losing future black voters and of protests that disrupt an orderly city. Thus, the discourse of racial transformation and threat of disruption remain the most powerful resources that the urban poor possess to counter neo-liberal managerialism.

Overall then, the challenge for democracy in Hout Bay is not one of repression or even depoliticisation, but mostly of the manipulation of participatory spaces in ways that reduce the responsiveness of the state. In part, this is because participatory spaces cannot compel elected officials to make decisions that are not in the interests of the elected official, the ruling party, or their core constituency. Notably, this unresponsiveness does come with a risk of raising the odds of protest and disruption to force accommodation of the urban poor, and officials do not want this either. Consequently, for the poor of Hout Bay much of politics occupies this ambiguous space between participation and disruption, whereas for decision-makers it lies between representation and wilful blindness.

Government and the limited power of City Hall

Most important to understanding the democratic deficit in Hout Bay and, we would suggest, the urban South more widely, is the limited power of elected officials over the multiple forms of governance that shape the City. This is an existing theme in the literature on the urban South, but that takes on much greater relevance when framed in terms of residents' experience of urban governance, rather than the powers of local government. This is because a citizen-centric view sidesteps the default assumption of Political Science that the state is the sole or most powerful actor in making rules and decisions on key events that affect citizens. A citizen-centric view means keeping this an open question. Of course, this is not to say the state is not important, far from it; indeed there is good evidence that the state is almost always involved in most forms of governing the City, but in different ways and to different degrees.

In exploring the power over the City of elected officials, we draw on Clarence Stone's (1989) contrast between power as control versus power as social production. We show how the power of elected officials is compromised both in terms of control over residents *and* in terms of the power to produce Hout Bay. The lack of control of elected officials of City Hall is manifested in three ways. The first is due to the division of powers between the spheres of government in South Africa. This means that City Hall lacks exclusive control over key state functions such as housing, education, health, and harbours. Furthermore, when any branch of government chooses to drive an agenda, for example when the Province decides to build a new school, they must work with the other relevant spheres in a co-operative fashion rather than as a subordinate in a hierarchy. Consequently, not only are there greater co-ordination challenges posed by this division of powers, but the process can only move at the speed of the slowest partner. As Western Cape Premier Helen Zille (2017) illustrates, 'I made upgrading the harbour in Hout Bay a priority project for eight years and got nowhere'. The main issues are the 'complexity of inter-governmental relationships' and the 'unresponsiveness of national departments'.

The second problem is the division of powers between elected and appointed officials. This is a problem related to the co-existence in contemporary democratic systems of both a Weberian bureaucratic hierarchy and the democratic accountability of elected officials. Simply put, while appointed officials must implement policies and decisions made collectively by the elected officials, they do not account to individual elected officials in implementing their tasks, but their line department hierarchies. Further, the bureaucracy is organised into line departments that operate as hierarchical silos, and officials do not account across their silos. This means that ward councillors and even subcouncils cannot instruct appointed local officials on what to do, especially when it comes to working against the hierarchical logics, mentalities, and scales of line departments. Consequently, innovations that require the City bureaucracy to work in unusual ways require direct instruction from the top elected officials only. Without it, there is no chance of innovation working, even when led by civil society organisations with community and local elected official backing, as illustrated in Chapter 2. These kinds of hierarchies of power between spheres of government, and between elected and appointed officials, are common across the world.

The third constraint on the power of elected officials is peculiar to the urban South, and concerns the significant degree of informality in settlement, livelihoods, and even governance. Informality is probably best understood in terms of the influx of poor migrants into the city who cannot afford to live by the rules of the formal system, but must live somewhere and secure a livelihood. Indeed, the rapid influx of poor migrants poses the greatest challenge to rule in Hout Bay, Cape Town, and across the urban South. In general, the response of City Hall is a pragmatic recognition that it lacks the capacity to fully include the urban poor, and thus it tolerates informality until the externalities become intolerable. In Hout Bay, these externalities have been moments of political crisis such as the Battle of Hangberg, economic crisis such as taxi violence, or natural crisis such as the fires that razed Dontse Yakhe to the ground. In response, the City incorporated some informal groups further, but not entirely, into the formal system through building more flats, formalising the bus system, and attempting super-blocking.

Given the systemic inability of City Hall to incorporate all of the urban poor into formal systems, this type of inclusion is driven by moments of exception. Recognising this, activists often try to create a crisis through disrupting the formal system, although typically only when really desperate because disruption risks a repressive response from the state or private security forces. Thus, while the tactic of disruption combined with the discourse of racial transformation is the greatest resource the urban poor possess in Cape Town, using it is far from simple. Given the limited resources of the state, the constraints of democratic rights and the law, and the potential costs of sustained resistance, most of the time City Hall tends to turn a blind eye to informality, and prefers negotiation and co-option to coercion.

This discussion brings us squarely to Stone's point about the limited capacity of City Hall to produce the urban on its own. Granted, under the DA, the City of Cape Town believes that other actors, principally business, should play a key role in constructing the City, but as the case of enduring informality reveals, even if it wanted to, City Hall would be unable to create Hout Bay on its own. Thus, while City Hall, and the state more widely, does have some capacity to provide housing for the urban poor, it cannot match the need in

Hout Bay. Furthermore, it is much, much slower and often less efficient than both market actors and the poor themselves. Similar points hold in respect of the built environment more widely (malls, shopping centres, entertainment complexes), livelihoods, transport, and security. This is all evidence of the limited power of elected officials to both control and produce Hout Bay on their own. In our view, this limited capacity to meet the challenge of inclusion of the urban poor brought about by rapid urbanisation constitutes the greatest challenge facing rule in the urban South. This means that how states partner to produce the city is crucial to the future of the urban South. Given the centrality of the urban South to meeting humankind's economic, social, and environmental challenges, this politics of producing the urban South will shape the future of the planet.

For a more compelling account of this we now turn to describing the complex and often-contradictory nature of rule in the urban South.

Rethinking power and governance in the urban South

Our citizen-centric approach to governance alerts us to forms of decision-making or governance over urban life that proceed beyond the exclusive control of elected officials of City Hall. Although avoiding state-centric assumptions about urban power, our approach still retains the connection to government, politics, and democracy through the notion of governance as decision-making over key rules and events in the City. This is important because it helps us refine the general observations that urban rule sits at the intersection of international, national, and local actors, and business, state, and the urban poor, into more specific descriptions of how decision-making proceeds on the very issues of concern to residents. In addition, this citizen-centric framing of governance allows for the differentiation of forms of rule by different coalitions of actors according to distinct logics or rationalities of rule – another generally observed point of rule in the urban South (Watson 2009). A citizen-centric framing of governance thus enables a detailed and emergent analysis of systems of decision-making, how they relate to familiar state institutions and other actors, including residents, their distinct logics, and the contingent outcome of their intersection on an issue, and in a time and place.

In relating governance to decision-making, the issue of power is raised, and it is here that we begin to build an emergent theory of urban governance. Thus, we show that to understand urban power we need to move beyond Lukes' (2005) conceptions of power as domination to include power as social production. Further, however, we argue that Stone's (1989) conception of power as social production needs to be expanded beyond a focus on law and resources (and thus business) to include issues of knowledge and legitimacy in the urban South. In respect of governance, we argue that a focus on practice and rules reveals that urban rules are a contingent constellation of multiple and contending forms of governance, which we term a 'topological' framework of urban governance. The particular constellations of multiple and contending forms of governance simultaneously order and disorder the suburbs of the city.

Urban power as power 'over' residents and 'power to' construct the city

As Stone (1989) has persuasively demonstrated, understanding urban power only in terms of 'power over' the laws and policies of the city is insufficient to explain how rule proceeds. For him, we need to think of power in productive terms, especially in terms of the resources of business that are necessary to create the city. While we are persuaded by this conception of productive power and the importance of resources, in our cases we also found that more was required to produce the city. In addition to the resort to coercive power through the law and the power conferred by resources, the production of Hout Bay has also required significant forms of knowledge on how to act, and popular legitimacy to act as well. Thus, whether building private security estates or public housing, social actors need to acquire multiple forms of legal, administrative, engineering, and environmental knowledge to proceed successfully. Similar points apply in respect of, for example, applying for taxi licences or buying into the Bus Rapid Transport System, or civil society organisations negotiating the city bureaucracy to help clean up the Disa River.

Furthermore, the issue of the legitimacy to make the city is well demonstrated by the opposition to the legally endorsed clearing of the informal settlement atop Hangberg, the resistance to super-blocking in Imizamo Yethu, and the sustained opposition to tolling Chapman's Peak Drive. Importantly, legitimacy is not only a question of popularity but also of legality, as revealed by the invocation of law to facilitate the densification of Hout Bay, despite the opposition of the HBRRA, and to suppress protests by disgruntled fisher folk or informal settlers. Having the power to produce the city thus requires aligning both legal rights and popular support to build a new Toll Road, clean the river, or construct a new school in Imizamo Yethu.

This enlarged conception of productive power implies greater potential variety in the forms of governance to produce the city. This is because the more variables that are involved, the greater the number of partners that might be of use in amassing productive power. Who these partners are, what they contribute, and in what proportion, could vary tremendously, opening the possibility of a great variety of forms of governance. This granted, the more the dimensions of productive power, the greater the likelihood that City Hall, and certainly the state more widely, will be involved in some way. This is because to each of the key elements – rules, resources, knowledge, and legitimacy – the state has something to contribute. Indeed, outside of pure forms of informal governance, it is necessary for the state to be involved as it retains exclusive power over the administration of law, and law is a key source of both the legitimacy and coercive capacity needed to govern. Framing urban power in terms of both power over residents, and the power to produce the city, opens up the conception of rule to include forms of decision-making centred on control, as well as forms of decision-making centred on social production. On both these aspects, local governments in the urban South cannot proceed alone.

Informality and the limits of formal control

As already noted, perhaps the distinctive feature of Southern urbanisms is the large populations of urban poor who live outside the rules of the formal system. This we have framed as a limit on the power of City Hall, and the state more widely, as it is simply impossible to coerce large groups of people to live by the formal rules when they do not have the capacity to do so. It is also important to note that it is not just City Hall or the state that lacks the capacity to control fully the urban poor, but all social actors in the urban landscape. Hence the significance of framing informality in terms of productive power, rather than just coercive or legal power. Addressing the challenge of the urban poor brought about by rapid urbanisation is primarily a problem of how to create the city in inclusive ways that address the big challenges of economic, social and environmental sustainability.

In addition to blurring the boundaries of formal control, informal life in the urban South manifests forms of governance that challenge the notion of governance being only about institutions. Hence, in identifying the important role of a criminal gang leader in deciding livelihood and security issues in Hangberg, or the contingency around who makes it onto a housing beneficiary list, we join a large body of literature on armed non-state actors who decide how certain residents live. The point of this observation is to draw attention to the reality that these actors may govern through more personalised, arbitrary, and less rule-bound ways. Thus, who gets punished for a transgression in an area controlled by drug gangs is often arbitrary and disproportionate. This personalisation of power, and the ways that formal rule is corrupted by informal practices, means there is more to governance in the urban South than rules and well-ordered institutions. In part, informality can be seen as about the disruption of formality to facilitate inclusion.

This granted, it is important not to overstate the 'unruliness' of informality for two reasons. The first is that much informality may take the form of corruption, nepotism or special favours by actors in formal governance, who break formal rules only some of the time. In this case informality consists in a 'greyness' about when and to what degree the formal rules are followed, rather than a lack of formal rule altogether. Second, even in those cases where urban life is governed outside of formal governance systems, informal rulers might institute basic sets of rules over daily life. Perhaps the best example of this is Arjona's (2014) account of wartime institutions, that demonstrates how, even in a war context, rebel groups may set up and enforce public rules over issues such as safety and security, health practices, transport, and the like.

In short, then, informality alerts us to the limits of control by elected officials over key practices of poor residents in the urban South, but also to the limits of control by elected officials over non-state leaders. In addition, it highlights that governance is not only about instituting rules, but also about the capacity to enforce ad hoc decisions too, often against the formal rules. In part, informality is a contestation of the urban order by the poor, and at times by criminal elements, a contestation that may involve the personalisation of power. Thus, informality challenges assumptions of urban rule as being only about institutions or well-ordered sets of rules typical of most Northern accounts of governance, including Foucault's. It is to these that we now turn.

FUG and the production of the City

A citizen-centric approach to urban rule in the Global South reveals not just a multiplicity of rulers, formal and informal, but also many forms of governance distinguishable in terms of rationality or logic. We identify at least five kinds in Hout Bay, although there may well be more, and we certainly need to explore in more depth the rules and rulers of informality. In thinking about governance in terms of rationalities, we follow Foucault's (1980) insight that power is better conceived of existing 'only in action' in social practices, and governance as constituting the forms of knowledge and power that condition these actions. This is an approach that complements our citizen-centric view in focusing on what urban residents actually do and experience, and the rules that inform this, rather than starting with formal institutions.

Similar to our account of urban power, Foucault's account of power is also both repressive and productive, and thus individuals are both 'subject to the constraints of the social relations of power ... and simultaneously enabled to take up the position of a subject in and through these constraints or operations of power' (Hamilton 2014: 76). Foucault is also instructive for our account of urban rule because, as Stephen Collier (2009) observes, he saw multiple forms of governance as co-existing in contingent combinations or 'topological' fashion. This means the simultaneous and contingent co-existence of divergent forms of governance in potentially changeable relationships. Furthermore, it is the combination of these forms of governance into an emergent totality or framework of urban governance (FUG), that defines the character of urban rule in a place.

Thus, in what follows we outline individually the five main forms of governance we identify in Hout Bay, before reflecting on the way they intersect or relate. In characterising them, we draw on Bevir's (2012) typology of forms of governance, but expand it to include actors and aims. In this context, 'actors' refers to the conception of 'who rules and who is ruled', and 'aims' refers to the fundamental objective that informs the logic of governance (see Table C.1).

Bureaucratic governance

Bureaucratic governance refers to the exercise of power by City Hall alone in passing by-laws, creating policies and making decisions, and in the routine implementation of these by the City's line departments. Thus, every day in

Table C.1 Main forms of governance in Hout Bay

Governance logics	Bureaucratic	Market	Developmental	Network	Informal forms (e.g. gang)
Actors	governors and governed	consumers and producers	managers and populations	stakeholders	bosses and residents
Aims	create rational order	efficient exchange	secure well-being for marginal groups	secure common interests	secure elite interests
Basis of relations	authority	free exchange	management	co-operation	command
Governed by	equal treatment	prices	developmental needs	trust	fear
Conflict resolution	rules and commands	law	rules and negotiation	negotiation	command
Culture	subordination	competition	consultation	reciprocity	subordination
Examples	refuse collection	security estate	flats in Hangberg	CPF Hout Bay	criminal bosses in Hangberg

Hout Bay the City of Cape Town can be seen collecting refuse, managing recycling, fixing potholes, measuring factory emissions, and patrolling the beach. Importantly, while city officials sometimes refer to residents as 'clients', the relationship is not one of supplier and consumer, as all residents are entitled to equal services, although they must follow the prescribed (often slow and complicated) processes to access these. Indeed, City Hall has put significant energy into feedback systems for residents in line department functions, in this way institutionalising responsiveness through what the World Bank (2004) describes as the 'short route' of accountability.

Furthermore, as illustrated in several places throughout the book, the institutions of formal democracy also influence line department operations. This is what the World Bank (2004) terms the 'long route' of accountability. Thus, the ward councillor and other politicians on the subcouncil advocate for the interests of their voters around municipal issues such as water, electricity, cleaning of the settlement and the river, and zoning regulations. To put this another way, formal local democratic institutions are connected to bureaucratic governance by City Hall, albeit in terms of the constraints of the formal hierarchies between elected officials and civil servants identified above.

Market governance

Market governance is co-governance between the state and market actors, most fully described in respect of the residential property market. There are many other forms too, notably the provision of private security, the use of private transport, and the mention made of private education. In all these cases market governance can be seen as a partnership between the national state that defines and upholds the system of property relations (including contract law through the courts), the local state (zoning and licensing), and private actors who then provide or purchase a service or product. Thus, where various spheres of the state have background 'power over' the conditions of the market through property and contract law, individuals and businesses have the 'power to' implement contracts for various goods and services.

Importantly, market governance of residential property has shaped *where* most people live in the City. It is private capital, rather than City Hall, that has set the agenda of middle-class and wealthy urbanisms by taking advantage of the national legal system of private property. In particular, private capital, including a significant number of international actors, has invested in residential property, especially in security estates and new malls, raising the value of all property in Hout Bay, even the shacks. That market governance of the property market is linked to globalisation of financial capital is widely observed (Davis 2006; Harvey 2012), and offers good reason to assume similar forms across the Global South (and North).

Perhaps the key consequence of the rise in land prices is that it makes access to land for poor people practically impossible, reproducing in class terms the racial exclusions from land of the apartheid era. High property prices are a key reason that the state looks to build infrastructure for Imizamo Yethu on the adjacent 16 hectare 'green belt', rather than more centrally in Hout Bay. The rising housing market and the unwillingness or inability of the state to expropriate land for reasons of social and spatial justice, mean that apartheid exclusions endure. Indeed, the opening of the land market to wealthy international buyers has further undermined the opportunity for poor, black South Africans to afford land. Similar logics are at play with fishing in Hangberg, where the combination of exclusion from licences by national policy and the incentives of global markets have turned a long-standing community of fishers into smugglers, poachers, and protesters.

In addition to shaping where people live, property developers shape *how* the middle classes and wealthy live by building lifestyle security estates, malls, shopping centres, and private schools all designed to enable more luxurious consumption by the well-off. Notably, security estates construct a well-ordered, safe and micro-managed neighbourhood not possible outside the high walls and electric fencing around private land. Indeed, this is a major selling point for investors. Outside the walls of security states, where many middle-class and wealthy people still live, civil society organisations, like neighbourhood watches, champion the vision of the City as a green, leafy, safe suburb for the well-to-do.

Lastly, the embracing of a Public-Private Partnership to toll Chapman's Peak Drive effectively authorised a private contractor to govern the use of a formerly public road with the backing of the law, while removing the requirement that it account to the public through politicians. Even the head of the Provincial government was prevented by law from holding the contractor accountable to residents. Indeed, it was only the resistance of local residents that delayed the building of the Toll Office in 2012, and so generated the incentives for the contractor to renegotiate the original contract. It is the state-sanctioned power of the legal contract that defines market governance, and can trump democratic accountability.

Developmental governance

Developmental governance refers to attempts by the state and/or donor organisations to improve the well-being of needy populations. Key here are infrastructure projects such as housing, schools, clinics, and sports fields, but also philanthropic and citizenship-building work in areas such as HIV/AIDS and health education. Three key points emerge from this Hout Bay case study. First, local government is seldom the only actor involved in development, and often not the lead actor. On key issues, provincial and national government departments usually lead. Second, developmental governance influences how poor people live, but seldom where they live. Third, developmental governance is framed as a partnership between the state/donors and residents, often mediated by civil society organisations, but residents are treated as populations of the needy to be managed for their well-being, rather than as individuals bearing rights. Further, in South Africa this biopolitical approach is framed in terms of the discourse of 'racial transformation' that gives need a racial character.

City Hall is not necessarily the driver of developmental governance in South Africa. This is the case as a consequence of a formal division of powers in the Constitution that gives the provincial and national spheres lead roles in key areas such as education, health, and housing. Indeed, while the metropolitan councils, such as the City of Cape Town, are assuming a greater role in developmental governance, most of the time projects are driven by other spheres of government, with City Hall involved as the partner with access to land (Zille 2017). When it works, this sharing of responsibilities is a classic instance of co-operative governance, but always within nationally developed policy frameworks. Thus, Provincial government took the lead in the Chapman's Peak Drive Toll Road, the Imizamo Yethu and Hangberg housing projects, the Disa School, and the new polyclinic. All of the time, though, developmental governance led by the state is a slow, cumbersome and difficult bureaucratic process, not least because of the challenges of working across both spheres and silos of government (Zille 2017).

Furthermore, developmental governance operates in terms of a biopolitical governmentality that focuses on uplifting needy populations, rather than meeting the rights of individual citizens. As Chatterjee (2004) observes, this is a common politics across the Global South, linked to the idea of the developmental role of the state. Chapter 6 provides a good example of this in the rebuilding of the Dontse Yakhe informal settlement after the 2017 fire. The approach of the City is to frame rebuilding ('super-blocking') as exclusively about providing decent shelter for poor residents. However, in providing everyone with the same 6m x 6m shack with water and electricity, the City is inadvertently removing some people's large shacks that they use for rental or informal livelihoods, such as alcohol retailing. Unintentionally, then, the focus on 'shelter' alone threatens innovative attempts by the poor to make a living. This inability to 'see' local practices, and thus inadvertently threaten them, is one reason for the resistance to upgrading.

Conversely, the logic of developmental governance incentivises residents to present themselves to the state as needy of resources. In South Africa, this intersects with the history of repression and exclusion to produce a racialised conception of 'needy groups', evident in discourses of transformation, Africanisation, and decolonisation. Furthermore, in a context of limited supply, development often gives rise to new forms of competition among 'the needy'. While SANCO tries to monopolise these development opportunities in Imizamo Yethu, this is hard to do in the absence of formal authorisation. Thus, development projects often prompt the local leadership conflicts associated with accessing state resources, for example in the struggle between Sinethemba and factions of SANCO over housing projects in Imizamo Yethu.

Network and informal forms of governance

Bureaucratic, market, and developmental governance were the three main forms we observed in Hout Bay, and there are excellent reasons linked to the globalisation of the modern state, neo-liberal capitalism, and international development discourses why we would expect to find these in some form across the urban South. However, they were not the only forms of governance we found in Hout Bay, and they are certainly not the only forms of governance that there could be. We hold this as always an open and empirical question. In our case though, two more are worthy of mention here: network and various informal forms of governance, one example being authoritarian rule by criminal bosses.

Network governance, called 'nodal' governance in respect of security governance (Shearing & Wood 2003), is a notable aspect of security provision. This involves the co-operation between civil society formations, specifically neighbourhood watches, the local police, and private security firms in patrols and surveillance, to prevent crime. These kinds of relationships are common across the suburbs and settlements of Cape Town, indeed the world, and they substantially improve security provision without involving City Hall at all. Of course, the state is involved in the form of the police, and in the supporting condition of the legal system, but a wide variety of other actors co-design and implement crime prevention strategies.

Informal forms of governance refer to those rules and programmes made outside or against the rule of the state, such as informal livelihoods and security provision. Good examples in Hout Bay would be the control of the criminal gangs over important livelihoods and trade in Hangberg and, arguably, the curfew imposed by the community patrols in Imizamo Yethu. Both sets of practices are illegal, but enforced by armed non-state actors, regardless. In the case of the Imizamo Yethu community patrols, the police give implicit support as they have had a positive impact on crime in the settlement, and thus help the state in achieving its objectives. Importantly, however, these moments of informal governance by armed non-state actors are a very small part of the informality story in Hout Bay. Indeed, informality is by definition a residual category, and more work is needed to identify in particular places the actual rules and rulers operating outside of the formal system.

Implications of FUG

Unpacking urban rule in Hout Bay from a citizen-centric perspective has revealed how residents encounter a wide range of authority figures, and several often-divergent forms of governance. Furthermore, these forms of governance may intersect in ways that reinforce or challenge democratic rule. In what follows, we reflect on this by exploring the implications of topological governance for political subjectivity, place, and democracy. In short, we show that citizenship is fragmented in time, governance is segregated by place, and democracy is disconnected from the multiple and contending forms of power that shape the City. While these features likely apply across much of the Global South, in South Africa they produce a form of socially, spatially, and politically divided demos we term 'neo-apartheid'.

Fragmenting citizenship: consumers, clients, and the marginalised

The promise of democratic rule is, in part, a promise of equal belonging and rights in a democratic state. This form of democratic citizenship exists in Hout Bay, but largely episodically, as democratic citizenship is practised only at particular moments in time and place: when in the voting booth during local or national elections; when in the community hall at a public meeting; when debating online or in the mass media; when protesting in the streets, and so on. For the vast majority of people these events are episodic, infrequent, and usually conducted in designated public spaces. Of course, there are residents who are not citizens of South Africa and therefore do not enjoy all the political rights offered to residents. In addition, there are inequalities in terms of who gets to speak in some public forums, linked to issues of party, gender, and nationality. Most significantly, there are public discourses of racial transformation, conservation and neo-liberal managerialism that question the right to belong in terms of nationality, race, party, and poverty.

However, more significant is the anti-democratic effect of multiple and contending forms of governance on democratic political subjectivity. This, following Wheeler (2014), we describe as 'fragmenting' citizenship. Thus, as the same time as living in Hout Bay as democratic citizens, most residents are also consumers of market governance, clients of the developmental state, or marginalised from formal rule, or some combination of the above. Were City Hall alone to rule, and to rule fully, the residents would access their daily needs through bureaucratic governance. In Cape Town, bureaucratic governance frames residents as 'clients to be serviced', very much the language of neo-liberalism but, notably, service is not monetised. In theory and mostly in practice, any resident can call the service centre and be served promptly. Thus, bureaucratic governance does attempt to treat all residents equally, and sees all residents as entitled to key services. Whatever its practical limitations as ponderous and complex, this form of governance is consistent with the promise of democratic rule: that all citizens belong and enjoy equal rights.

Market governance, by contrast, frames citizens as consumers and/or producers of goods or services. Indeed, this form of rule reduces citizenship to issues of consumption to the extent that national citizenship is almost never a condition for accessing the housing market in Hout Bay. Consumer citizenship is also formal and protected under law, in particular property, labour, and contract law, and looks to privatise the meeting of key needs from the state to business, and to depoliticise it from collective debate to individual grievance. Consumer citizenship even extends to privatising neighbourhood governance within gated communities, and the internal governance of security estates likely approaches Foucault's account of disciplinary power.

Developmental governance in turn imagines residents as 'needy' populations in relation to the global categories of human development, such as poverty, health, education, shelter, and the like. While these biopolitical categories can also be framed in terms of rights, in South Africa they tend to be deployed collectively to racialised groups rather than individuals, and are conjoined in the idea of 'racial transformation'. By framing residents collectively and in terms of well-being, residents are no longer individual citizens with rights so much as racialised groups of over- or under-privilege. In addition, in the South African context, access to the state is often mediated by partisan actors, who capture political office for career purposes as professional mediators or gatekeepers (Piper & Bénit Gbaffou 2014; Piper 2015; Piper & Anciano 2016). Hence, the frequent complaints of corruption, nepotism and political clientelism in respect of developmental governance.

Lastly, informality involves forms of agency that occur outside the legal or policy framework of formal rule, but require management by the state. This places these residents in an ambiguous position of living outside the formal and thus being unable to access rights fully, but also unable to be fully ignored. For example, living in an informal settlement places residents outside of the possibility of consumer citizenship as legal buyers or renters, and while they remain objects for developmental governance, must live illegally, waiting for the state to deliver. While some argue that the foreigner lives in a state of pure exclusion from local rule, this is not entirely the case. Not only do refugees enjoy almost all the social and economic rights South Africans have, but the urban poor also exist in a similar position of illegality to illegal immigrants in respect of meeting daily needs. Both groups reside, commute, socialise, and secure income to a significant degree outside the law. While there are clearly degrees of marginality, this category of resident exists in the grey zone between legality and criminality.

Consequently, on a quotidian basis, democratic citizenship is the exception rather than the rule of governance in Hout Bay. It is found episodically and almost always in designated public places. More common is the pursuit of the needs of daily life, often framed outside of individual rights-bearing citizenship and in terms of market, developmental, and informal rules and rulers. Most of the time, then, the residents of Hout Bay are constructed by governance systems as consumers, clients, or marginal groups, and only some of the time are they democratic citizens. Notably, these forms of political subjectivity threaten to continue the marginality of black residents of the City, long denied democratic citizenship under apartheid on the grounds of race.

Importantly, as Gaventa (2010) notes, in a democratic context this diversity of forms of political subjectivity also offers opportunities for residents to assert themselves in ways not imagined by powerful institutions and actors. Rather than reducing political subjectivity to logics of governance, we need to remain open to the agency of residents in creatively using the varieties of opportunities that confront them, and in constructing new notions of belonging, being, and politics in this process. A good example of this from Hout Bay would be how refugee rights organisations contest both popular prejudice and formal exclusion in the language of democratic rights and a shared humanity.

Segregating place: the green, grey, and brown of neo-apartheid

In Chapter 3 we note how the enduring racial segregation of the apartheid era is reproduced today in Hout Bay through economic means. Due to its enduringly racial nature, we term this combination of historical and economic segregation 'neo-apartheid'. Importantly, there is also a political character to neo-apartheid segregation due to the different frameworks of governance manifest in the major settlements of Hout Bay, even across the same issues of housing, education, or transport. These manifestations result in differing degrees of coherence in governance and order across Hout Bay.

The topological nature of urban governance means that different frameworks of the several forms of governance coexist in different parts of the City. For example, the developmental governance of housing may coexist alongside market and informal forms in the City as a whole. A consequence of this is a significant degree of inconsistency of rule in Hout Bay. For example, building codes apply in the Valley but barely in Imizamo Yethu, and unevenly in Hangberg. Similar points can also be made about drinking in public or personal security on the streets of the settlements. Key to understanding this incoherence is the limit of formal rule identified above, but also the fact that different forms of governance operate in terms of distinct logics. To characterise and explore this, it is useful to reflect on how the same issues are governed differently in the Valley, Imizamo Yethu, and Hangberg respectively.

In the Valley, many of the same processes that explain access to property in Hout Bay apply in respect of the governance of other issues too. Thus, wealthy people who can afford private housing also tend to be privately employed and purchase their transport, education, health, and security needs from the market, too – or more accurately 'top up' the transport, education, health, and security provision of the state. Indeed, market governance is by far the most common form in the Valley, and complemented by bureaucratic governance such as water and electricity provision, road maintenance, and cleaning services. On the whole then, the Valley can be described in Murray's (2015) terms as a 'post-public' wealthy enclave, where the governance logics of market and state largely cohere, and the area is well ordered with a set of well-known and extensive rules that are generally observed. We term these kinds of areas 'green' to reflect the primacy of money, but also the ideas of conservation and environmentalism propagated by organisations representing the Valley.

Neither Hangberg nor Imizamo Yethu are green areas. The key reason is that both of these settlements have limited market governance; most residents are not wealthy enough to buy private transport, education, health-care, or security, instead relying on the state to provide it, or doing it themselves. Given the significant number of poor residents in both areas, this is where most developmental governance in Hout Bay occurs. Importantly, both areas, but especially Imizamo Yethu, have significant informal practices including people building their own shacks and trading, securing livelihoods, visiting traditional healers, and forming informal street patrols. Both are thus what Yiftachel (2009) terms 'grey areas', where many residents live across or between formal and informal, legal and illegal binaries.

Notably, rule in 'grey' areas such as Hangberg and Imizamo Yethu is both contradictory and less ordered than that in 'green' areas such as the Valley. Part of the reason for this is that contradictions exist between the logics of market, bureaucratic, developmental, and informal governance. For example, state housing is not related to the formal market as beneficiaries cannot sell for 8 years, and are unable to raise a bond or loan from a bank against the house. At the same time, state housing is designed to accommodate a specific number of people per electricity connection or per room, but these resources are quickly stretched through the rental of rooms and backyard spaces to informal settlers. In practice, beneficiaries of developmental governance face incentives to treat their house as a financial asset on the informal market. Lastly, the extent of informality means that groups and individuals behave in ways beyond the formal rules, some of which may be informal, but much of the time are simply less extensive.

In many ways, the pressure on democratic citizenship in Hout Bay is symbolised spatially through the loss or exclusion of public space, by way of the construction of green and grey spaces. Where private governance fences out the public, informality crowds out public spaces, appropriating them for housing or other non-public needs. These divergent urbanisms thus erode the places for a shared, democratic citizenship in favour of green consumerism or grey developmentalism. Democratic place shrinks to the form of the public school, library or community hall. Last, but not least, Hout Bay is unusual in Cape Town in that it has very little in the way of a formalised working-class/ lower-middle-class suburb or 'brown area' that falls between 'green' and 'grey' areas, and that make up much of Cape Town.

Local democracy disconnected

Hopefully by now it is clear that exploring popular dissatisfaction with democratic rule requires framing the problem of local government as one of local governance. Understood as forms of decision-making over rules and key events in the City, the governance approach enables us to unpack the multiplicity of actors who shape urban life, their respective power, and how that power is exercised in controlling residents or creating the City. Identifying five main forms of governance (bureaucratic, market, developmental, nodal, and informal) across the settlements of Hout Bay, we show how these forms coexist in ways that produce well-ordered 'green' spaces for the wealthy, mostly ordered 'brown' spaces for working people, and unevenly ordered 'grey spaces' for poor and marginal residents. Partly, this is due to the contradictory logics of forms of governance in brown and grey spaces, but also to the limits of formal rule altogether for poor and migrant groups unable to live by the rules of the City. The segregation of place in economic and social terms thus has a political correlate in governance terms too.

The motto of the City of Cape Town is 'The city that works for you'. This is a sentiment we would largely share in respect of the governance of roads, refuse, waste, electricity, and so on. However, urban rule is more than bureaucratic governance by City Hall, and the areas in which the wealthy live are the most coherently governed and well ordered, while the areas in which poor people live are more contested, chaotic, and informal. While it is tempting to read into this governance outcome the notion that City Hall rules for the wealthy and not the poor, this makes the mistake of assuming that City Hall governs alone. This is simply not the case, especially as rapid urbanisation poses an unprecedented challenge for the production of the city. We thus still need to explore the extent to which an informal coalition of some sort currently rules Cape Town in the interests of the wealthy, or whether the current urban order is a more contingent outcome of multiple unco-ordinated actors and processes. This is a question that requires more research at the level of City Hall politics and policy, rather than citizens' experiences of governance.

In terms of democratic rule, our account helps show why spaces for public life are shrinking, why opportunities for democratic citizenship are challenged by forms of governance that imagine consumers, clients and ignore the marginalised, and why opportunities to secure urban needs are highly shaped by wealth and preferential state access. However, what it reveals most clearly is the profound disconnect of local democracy from the key forms of governance that shape urban life. Simply put, the elected officials of City Hall have no, or only episodic, influence over informality, little influence over market governance, some influence over certain developmental projects, and the greatest influence over maintaining roads, collecting refuse, and supplying water and electricity.

To make a more profound difference in the lives of its residents, the City needs to partner with provincial and national spheres of the state, business, and civil society to develop the harbour, create tourism jobs, improve health-care provision and quality education, clean up the river and the air, and enhance safety and security. In short, to give residents what they want, the institutions of local democracy need to be *connected* to the forms of governance that actually produce and sustain the City. At the moment, democratic institutions promise what City Hall cannot deliver. In conclusion then, the democratic deficit of the urban South is rooted in the emergent disjuncture between the promise of participation, and the lived experience of remote governance.

References

Arjona, A. (2014). Wartime Institutions: A Research Agenda. *Journal of Conflict Resolution*, *58*(8), 1360–1389.

Bevir, M. (2012). *Governance: A Very Short Introduction*. Oxford: Oxford University Press.

Chatterjee, P. (2004). *The Politics of the Governed: Popular Politics in Most of the World*. New York: Columbia.

Collier, S. J. (2009). Topologies of Power: Foucault's Analysis of Political Government Beyond 'Governmentality'. *Theory, Culture & Society*, *26*(6), 78–108.

Davis, M. (2006). *Planet of Slums*. London: Verso.

Foucault, M. (1980). *Power/Knowledge. Selected Interviews and Other Writings, 1972–77*. Brighton: Harvester.

Gaventa, J. (2010). Seeing like a Citizen: Re-Claiming Citizenship in the Neoliberal World. In Fowler, A., & Malunga, C. (Eds.), *NGO Management: The Earthscan Companion*. London: Earthscan.

Hamilton, L. (2014). *Freedom Is Power: Liberty Through Political Representation*. Cambridge: Cambridge University Press.

Harvey, D. (2012). *Rebel Cities: From the Right to the City to the Urban Revolution*. London: Verso Books.

Ledeneva, A. (Ed.). (2018). *The Global Encyclopaedia of Informality, Volume 2*. London: UCL Press. www.ucl.ac.uk/ucl-press/browse-books/global-encyclopaedia-of-informality-ii. Accessed 26 March 2018.

Lukes, S. (2005). *Power. A Radical View*. Second Edition. Basingstoke and New York: Palgrave MacMillan.

McClendon, G. H. (2016). Race and Responsiveness: An Experiment with South African Politicians. *Journal of Experimental Political Science*, *3*(1), 60–74.

Murray, M. (2015). City of Layers: The Making and Shaping of Affluent Johannesburg after Apartheid. In Haferburg, C., & Huchzermeyer, M. (Eds.), *Urban Governance in Post-Apartheid Cities: Modes of Engagement in South Africa's Metropoles*. Pietermaritzburg: University of KwaZulu-Natal Press, 179–198.

Norris, P. (2011). *Democratic Deficit: Critical Citizens Revisited*. Cambridge: Cambridge University Press.

Piper, L. (2015). From Party-State to Party-Society in South Africa: SANCO and the Informal Politics of Community Representation in Imizamo Yethu, Hout Bay, Cape Town. In Bénit Gbaffou, C. (Ed.), *Popular Politics in South African Cities: Unpacking Community Participation*. Pretoria: HSRC Press, 21–41.

Piper, L., & Anciano, F. (2015). Party over Outsiders, Centre over Branch: How ANC Dominance Works at the Community Level in South Africa. *Transformation: Critical Perspectives on Southern Africa*, *87*, 72–94.

Piper, L., & Bénit Gbaffou, C. (2014). Mediation and the Contradictions of Representing the Urban Poor in South Africa: The Case of SANCO Leaders in Imizamo Yethu in Cape Town, South Africa. In von Lieres, B., & Piper, L. (Eds.), *Mediated Citizenship: The Informal Politics of Speaking for Citizens in the Global South*. Basingstoke: Palgrave Macmillan, 25–42.

Piper, L., von Lieres, B., & Anciano, F. (2017). The Tale of Two Publics: Media, Political Representation and Citizenship in Hout Bay, Cape Town. In Garman, A., & Wasserman, H. (Eds.), *Media and Citizenship: Between Marginalisation and Participation*. Pretoria: HSRC Press, 120–138.

Shearing, C., & Wood, J. (2003). Nodal governance, democracy, and the new 'denizens'. *Journal of Law and Society*, *30*(3), 400–419.

Stone, C. (1989). *Regime Politics: Governing Atlanta, 1946–1988*. Lawrence, KS: University of Kansas Press.

Wacquant, L. (2010). Crafting the Neoliberal State: Workfare, Prisonfare, and Social Insecurity. *Sociological Forum*, *25*(2), 197–220.

Watson, V. (2009). Seeing from the South: Refocusing Urban Planning on the Globe's Central Urban Issues. *Urban Studies*, *46*(11), 2259–2275.

Wheeler, J. (2014). 'Parallel Power' in Rio de Janeiro: Coercive Mediators between Citizens and the State. In von Lieres, B., and Piper, L. (Eds.), *Mediated Citizenship: The Informal Politics of Speaking for Citizens in the Global South*. Basingstoke: Palgrave Macmillan, 72–92.

World Bank. (2004). *World Development Report: Making Services Work for Poor People*. The International Bank for Reconstruction and Development. https://openknowledge.worldbank.org/bitstream/handle/10986/5986/WDR%202004%20-%20English.pdf?sequence=1. Accessed 28 October 2016.

Yiftachel, O. (2009). 'Critical Theory and 'Grey Space': Mobilization of the Colonized'. *City*, *13*(2–3), 246–263.

Interview

1. Zille, H. (2017). Premier of the Western Cape. Interviewed by Laurence Piper. 26 September 2017.

Index

Please note that page references to figures will be in **bold**, while references to tables are in *italics*. Footnotes will be denoted by the letter 'n' and note number following the page number. 'CCT' stands for 'City of Cape Town'.